Climate change
and the oil industry

Issues in Environmental Politics

Series editors Mikael Skou Andersen and Duncan Liefferink

At the start of the twenty-first century, the environment has come to stay as a central concern of global politics. This series takes key problems for environmental policy and examines the politics behind their cause and possible resolution. Accessible and eloquent, the books make available for a non-specialist readership some of the best research and most provocative thinking on humanity's relationship with the planet.

already published in the series

Science and politics in international environmental regimes
Steinar Andresen, Tora Skodvin, Arild Underdal and Jørgen Wettestad

Congress and air pollution: environmental politics in the US
Christopher J. Bailey

Implementing international environmental agreements in Russia
Geir Hønneland and Anne-Kristin Jørgensen

The protest business? Mobilizing campaign groups
Grant Jordan and William Maloney

Implementing EU environmental policy
Christoph Knill and Andrea Lenschow (eds)

Environmental pressure groups Peter Rawcliffe

North Sea cooperation: linking international and domestic pollution control
Jon Birger Skjærseth

European environmental policy: the pioneers
Mikael Skou Andersen and Duncan Liefferink (eds)

Environmental policy-making in Britain, Germany and the European Union
Rüdiger K. W. Wurzel

Climate change
and the oil industry

Common problem, varying strategies

Jon Birger Skjærseth and Tora Skodvin

Manchester University Press

Manchester and New York

distributed exclusively in the USA by Palgrave

Published by Manchester University Press
Oxford Road, Manchester M13 9NR, UK
and Room 400, 175 Fifth Avenue, New York, NY 10010, USA
www.manchesteruniversitypress.co.uk

Distributed exclusively in the USA by
Palgrave, 175 Fifth Avenue, New York NY 10010, USA

Distributed exclusively in Canada by
UBC Press, University of British Columbia, 2029 West Mall,
Vancouver, BC, Canada V6T 1Z2

British Library Cataloguing-in-Publication Data
A catalogue record for this book is available from the British Library

Library of Congress Cataloging-in-Publication Data
A catalog record for this book is available from the Library of Congress

ISBN 13: 978 0 7190 6559 0

First published in hardback 2003 by Manchester University Press
This paperback edition first published 2009

Printed by Lightning Source

Contents

List of tables and figures   *page* vi
Preface   vii
Acronyms and abbreviations   ix

1 Introduction   1
2 Analytical framework   12
3 The climate strategies of the oil industry   43
4 The Corporate Actor model   74
5 The Domestic Politics model   104
6 The International Regime model   158
7 Concluding remarks   196

Appendix: personal communication   221
References   223
Index   237

List of tables and figures

Tables

3.1 Summary of the climate strategies of ExxonMobil, the Shell Group and Statoil   *page* 69

4.1 Oil and gas reserves in 2000: ExxonMobil, Shell and Statoil   77

4.2 The CA model: expected versus actual strategies in relative terms   94

5.1 Percentage expressing 'a great deal' of concern in the US   109

5.2 Improvement of energy efficiency in Dutch oil and gas production   126

5.3 The DP model: expected versus actual strategies in relative terms   148

Figures

3.1 ExxonMobil's Operations Integrity Management System   47

4.1 Company structure: the Shell Group   89

4.2 Company structure: ExxonMobil   91

4.3 Company structure: Statoil   92

Preface

Climate change and a number of other environmental problems are partly, and sometimes mostly, caused by the legitimate activities of large corporations. Corporations, therefore, often control the behaviour that needs to be changed to solve the problem in question. The conventional way of studying the effectiveness of international environmental regimes or domestic environmental policy is to analyse the chain of consequences flowing *from* policy decisions *to* the strategies and behaviour of target groups. In this study, we have turned this research approach upside down by taking non-state target groups, i.e. large corporations, as our point of departure. Instead of starting out with joint international commitments or national policy goals, we have focused on corporate climate strategies. Why do corporations apparently operating within very similar business contexts nevertheless choose different strategies to confront a common problem? Which conditions trigger changes in corporate strategies?

The consequences of this 'bottom-up' approach immediately became clear when we planned this study: fact-finding had to start at the corporate level rather than at political level. In two rounds in March and November in 2000, we visited oil companies in the US and Europe. We also talked with representatives of the environmental movement, government authorities and various business organisations. To us as political scientists, the business community represented a new challenge, and our understanding of the sources of corporate strategy choice is coloured by our profession.

We started our work on this study in 1999. Preliminary findings were first presented at a side-event at COP-6 in The Hague in November 2000, and subsequently at a number of workshops and conferences. The positive feedback we received encouraged us to write a book on this topic. We got funding from the Norwegian Research Council's PETROPOL programme. This book would never have materialised, however, without additional financial support from the Fridtjof Nansen Institute, Center for International Climate and Environmental Research, Oslo (CICERO), and Department of Political Science at the University of Oslo. Lynn A. Parker Nygaard's help was invaluable. She helped us to make our arguments as logical and consistent as possible, and she provided language editing. Atle Christer Christiansen participated as a fact-finder in the project at an early stage. We have also benefited from his useful comments to earlier drafts. Many scholars have helped us with comments and valuable suggestions for improvements to earlier drafts. We would particularly like to thank Arild Underdal, Jørgen Wettestad, Per Ove Eikeland, Steinar Andresen, Jon Hovi, Asbjørn Torvanger, Asbjørn Aaheim, Dag Harald Claes, Knut Alfsen and Audun Ruud. Last, but not least, we would like to thank representatives of the oil companies, environmental movement and governmental organisations for taking the time to talk with us despite their no doubt tight time schedules.

The responsibility for any errors or misinterpretations rests with the authors.

Jon Birger Skjærseth, Lysaker
Tora Skodvin, Oslo

Acronyms and abbreviations

AGBM	Ad Hoc Group on the Berlin Mandate
AOSIS	Alliance of Small Island States
API	American Petroleum Institute
BA	benchmark agreement
BAT	best available technology
BP	British Petroleum
BTU	British thermal unit
CA	Corporate Actor
CAN	Climate Action Network
CCAP	Climate Change Action Plan
CDM	clean development mechanism
CHP	co-generation of heat and power
CNE	Climate Network Europe
CO_2	carbon dioxide
COP	Conference of the Parties
DoE	Department of Energy
DP	Domestic Politics
ECCP	European Climate Change Programme
EEB	European Environmental Bureau
EMAS	Eco-Management and Audit Scheme (European Union)
ENGO	environmental non-governmental organisation
EPA	Environmental Protection Agency
EUROPIA	European Petroleum Industry Association
FIELD	Foundation for International Law and Development
FoE	Friends of the Earth
GCC	Global Climate Coalition
GHG	greenhouse gas
HSE	health, safety and environment
ICC	International Chamber of Commerce

IEA	International Energy Agency
IGO	intergovernmental organisation
INC	Intergovernmental Negotiating Committee
IPCC	Intergovernmental Panel on Climate Change
IPIECA	International Petroleum Industry Environmental Conservation Association
IR	International Regime
LTA	long-term agreement
N_2O	nitrous oxide
NEPP	National Environmental Policy Plan
NGL	natural gas liquids
NGO	non-governmental organisation
NHO	Norwegian Employers' Association
NMVOC	non-methane volatile organic compounds
NO_x	nitrogen oxide
OECD	Organisation for Economic Cooperation and Development
OIMS	Operations Integrity Management System
R&D	Research and Development
SDFI	state's direct financial interest (Norway, on the Norwegian continental shelf)
SFT	State Pollution Control Authority (Norway)
SO_2	sulphur dioxide
STEPS	Shell Tradable Emission Permit System
UNCED	United Nations Conference on Environment and Development
UNEP	United Nations Environment Programme
UNFCCC	United Nations Framework Convention on Climate Change
UNICE	Union of Industrial and Employers' Confederation of Europe
USCAR	United States Climate Action Report
VNO	Netherlands Employers' Association
VOC	volatile organic compounds
VROM	Ministry of Housing, Spatial Planning and the Environment (Netherlands)
WBCSD	World Business Council for Sustainable Development
WGs	working groups
WMO	World Meteorological Organisation
WRI	World Resources Institute
WTO	World Trade Organisation

1

Introduction

In the prelude to the 1992 United Nations Framework Convention on Climate Change (UNFCCC), the oil industry was united in its opposition to binding climate targets. All major oil companies took the position that action on global warming could be damaging to their economic interests since the oil industry earns its livelihood from oil, gas and coal – the main sources of emissions of greenhouse gases. Ten years later, the positions of many oil companies have changed completely. Major European multinational oil companies such as BP (British Petroleum) and Shell support the Kyoto Protocol, have set ambitious goals to reduce their own greenhouse gas (GHG) emissions, and have invested in renewable energy. At present, these companies increasingly see themselves as energy companies rather than merely oil companies. Conversely, a major US-based company such as ExxonMobil – the biggest company in the world – has not changed at all. ExxonMobil opposes the Kyoto Protocol, it has not set any reduction targets for its own GHG emissions, and it does not have any immediate plans to invest in renewable energy.

The oil industry will be severely affected by regulatory measures to curb GHG emissions. With its multinational companies linked in worldwide operations, the oil industry constitutes a global industry operating in a global market. The business opportunities and challenges offered by the problem of climate change would thus appear to be the same for all large oil companies. This implies that the climate strategies of each oil company should also be the same. As stated above, however, this is not the case. The significant changes and differences apparently witnessed in the climate strategies of major oil companies thus represent an interesting puzzle.

The aim of this book is to shed light on this puzzle by exploring the extent to which and why major oil companies have adopted different strategies to address the climate issue, and by focusing on the conditions that have triggered changes in corporate strategies over time. How different are the climate strategies adopted by major oil companies on the climate change issue? Do the companies merely use different rhetoric or are the differences substantial? Why do the strategies of the oil majors vary and change over time, and what conditions trigger such changes? While interesting in their own right, these questions are also important for the prospects of establishing a viable international climate policy. Large multinational oil companies represent significant target groups for mitigating climate change. More than 50 per cent of GHG emissions originate from the activities of multinational corporations, and oil is responsible for about one quarter of the 'greenhouse effect' (Gleckman, 1995). Large oil companies influence domestic climate policy, affect the positions of states in international climate negotiations, and constitute critical target groups when policies are to be implemented. Against this backdrop, the identification of conditions determining the climate strategies chosen by the oil industry will provide knowledge about whether and how corporate resistance to a viable climate policy can be overcome.

There are essentially two main views regarding the extent to which large multinational corporations are controllable or not within the present world system. These views have important consequences for environmental governance. On the one hand, some argue that large corporations increasingly operate beyond political control. On the other hand, others dispute the validity of the claim that international economic integration or globalisation has produced the 'global corporation', which owes allegiance to no state. Admittedly, this book cannot settle this dispute, but it can contribute to the general understanding of the corporate scope of influence within the issue area of environmental policy and climate change.

Few would dispute the increasing importance of corporations, controllable or not, in the world economy as well as in environmental policy. Fifty-one of the largest 100 economies in the world (excluding banking and financial institutions) are now corporations (Retallack, 2000). Sales and assets of large multinational corporations far exceed the GNP of most countries in the world. Multinational companies are involved in 70 per cent of world trade and hold 90 per cent of all technology and product patents (Gleckman, 1995). In addition to climate change, a wide range of global environmental problems – including ozone depletion, loss of biodiversity, over-fishing, and illegal trade in hazardous wastes – has been linked to the worldwide activities of multinational corporations (Retallack, 2000). Corporations thus represent powerful forces in environmental degradation and international as well as national environmental policy (Rondinelli and Berry, 2000).

Despite the important role played by the oil companies in particular and large corporations in general, the primary focus of most academic climate policy and environmental studies has been on the robustness of science and the development and operation of national and international institutions, in which states and governments, the scientific community and the green movement have been pinpointed as the key players. Systematic comparative studies of multinational and even global multinational companies have been in short supply. However, there are some bits and pieces of relevant literature that can guide our analysis. In the next section, we shall present some of the main contributions.

The study of corporate actors in environmental policy

During the last decade, scholarly attention to the relationship between corporations and environmental politics has increased significantly. First, there is an emerging body of literature within the field of business management that focuses on how businesses deal with the impact of their activities on the natural environment (henceforth referred to as the business environmental management literature), which is helpful in addressing our first research question concerning differences in corporate strategic approaches to the climate issue. At a general level, this body of literature focuses on how corporate strategies can be categorised and how company-specific features determine corporate strategy choice (see, inter alia, Post and Altman, 1992; Roome, 1992; Steger, 1993; Ketola, 1993; Hass, 1996; Ghobadian et al., 1998). A

common thread throughout this literature has been a classifica-
tion of corporate strategy according to which companies are classified as being 'reactive'/'defensive', 'indifferent', 'proactive'/
'offensive', and 'innovative' (see, for instance, Steger, 1993). The
approaches developed within this school of thought shed light on
the 'greening of industry', and on why insurance companies are
more eager than car manufacturers or oil companies to reduce
long-term risks resulting from weather-related disasters
(Paterson, 1999). This literature has, however, less to say when
companies operating within the same branch adopt significantly
different strategies when faced with the same problem.

The business environmental management literature is also
useful in that it tends to relate different strategies to the same
behavioural mechanism: company survival and profits. Crudely
put, differences in strategies occur because there are many ways
to make money. However, some maintain that large multinational corporations share the same basic objectives of expansion,
growth and profit maximisation, paying only lip service to social
responsibility and environmental protection (e.g. Retallack,
2000).

The question of identifying the driving forces behind corporate
environmental strategies has recently been approached from a
variety of angles within several disciplines and schools of thought.
This literature tends to be organised around causal pathways
and/or different levels of analysis. Levy and Newell (2000) argue
that socio-cultural, political-institutional and corporate-strategic
factors explain differences and similarities between European and
US-based industry positions on ozone depletion, climate change
and genetically modified foods. Levy and Newell's main conclusions are that economic and competitive considerations appear to
dominate and that there is a trend towards convergence between
company strategies on either side of the Atlantic. Another case
study of European and American auto industries' responses to
climate change supports these observations (Levy and Rothenberg,
1999).[1] In his study of BP's and ExxonMobil's positions on
global climate change, Rowlands (2000) places more emphasis on
company-specific features. In this case, however, neither specific
market conditions related to the fossil-fuel portfolio, share of
operation in developing countries not committed to the Kyoto
Protocol, nor renewable-energy activities provided sufficiently

valid explanations of differences between these companies. Rowlands concludes by pointing to the need for further investigations of the importance of management structures and nationality, i.e. the home-base countries of corporations. Rowlands's observations have later been further explored and partly supported in other studies (Skodvin and Skjærseth, 2001; Skjærseth and Skodvin, 2001; Kolk and Levy, 2001).

While company-specific factors and home-base countries have received increased attention, the attention paid to the link between international institutions and corporations has been scant. Prominent international regime scholars have repeatedly emphasised the role and influence of non-state actors in international environmental policy (Levy et al., 1995). Corporations are particularly important to the analysis of regime effectiveness since industry is a major cause of environmental problems and thus represents a crucial target group. International regimes as well as governments depend upon the cooperation of corporate actors, whether active or reluctant, when adopting and implementing joint international commitments. In essence, a regime cannot be effective unless it is able to change the strategies and behaviour of relevant target groups.[2] Nevertheless, the analysis of international environmental regimes still tends to be state-centric (see e.g. Newell, 2000: 23; Miles et al., 2001). Neo-institutionalism, which in some version or another drives most analysis of international environmental politics, emphasises the importance of regimes, but downplays the role of non-state actors. Conversely, political economy emphasises non-state actors, but not regimes, while neo-realism downplays the influence and role of both regimes and non-state actors. Thus, Arts (2000) argues that relevant theories emphasise either regimes or non-state actors, or neither regimes nor non-state actors.[3] A number of studies, however, show that non-state actors frequently make a difference in international cooperation.[4] The roles of environmental non-governmental organisations (ENGOs) and the scientific community have received increased attention, while the role of companies in international environmental politics has, until very recently, been neglected.

There are, however, some valuable contributions on corporations and international environmental politics. First, there are some studies focusing on the role of large companies in global environmental governance (see e.g. Gleckman, 1995; Falkner, 1996). Emphasis is here placed on the power of companies to affect the state of the environment, as well as the economic and political forces and channels in which companies affect the establishment of global environmental governance in general. Previous efforts have also been made at a conceptual level to combine regime theory with non-state actors. Haufler (1993), for example, distinguishes between two types of non-governmental organisation (NGO), state and regime relationships: first, an instrumental relationship whereby the state is dominant, and a second type whereby non-state actors form independent regimes or play an equal role with states.

Second, the role of industry has perhaps been most intensively studied in the case of ozone depletion. The role of the chemical industry in negotiating the 1987 Montreal Protocol on ozone depletion has been studied in detail (see e.g. Benedick, 1991; Haas, 1991; Skjærseth, 1992; Maxwell and Weiner, 1993; Levy, 1997). On the one hand, the ozone case has similarities to the climate case in the sense that European and US-based chemical companies were split on the issue. In sharp contrast to climate change and the oil industry, however, the battle between the major producers – Du Pont in the USA and ICI in the UK – concerned markets for available substitutes for CFCs and halons. The chemical companies were not heavily dependent upon CFC production, and CFCs were not, in contrast to energy, critical to the modern industrial economy. In essence, the chemical industry faced a significantly more benign challenge than the oil industry.

Third, there are some recent studies focusing on companies and other non-state actors in international climate change policies. One approach highlights the channels of power through which multinational companies can affect regulatory international institutions. A distinction is here made between discursive, instrumental and structural influence (Levy and Egan, 1998). This topic has also been approached from the perspective of the influence of non-state actors in different phases of international climate cooperation: from agenda setting via negotiation to implementation (Newell, 2000). For a general analysis of the climate change negotiations emphasising the role of industry, see Leggett (1999).

This brief overview of some of the literature on corporations and the environment reveals a recently growing academic interest in these matters. Much of the appreciable literature referred to above has in fact been published during the course of this work and we will, as far as possible, seek to base our analysis on previous findings in order to facilitate cumulative knowledge. These studies, however, are also characterised by a number of shortcomings. First, the analyses are at a very general level. While there are some contributions that focus explicitly on differences between companies, a general focus on the link between corporations and environmental politics is still dominant. Second, this body of literature has to only a very small extent (if at all) focused on changes in corporate strategies over time. Third, it is to only a very small extent based on structured and systematic empirical analyses. Thus, there is clearly a need for developing a more comprehensive analytical framework aimed at systematic empirical scrutiny of corporate strategies to address environmental issues.

Research strategy

The analysis aims at both a synchronic comparison of corporate strategy choice and a diachronic comparison of changes in corporate strategies over time. The focus for explanation in this study is thus differences and change in corporate climate strategies. Conceptualising and measuring such differences within the same branch is by no means a simple task. The public profile of an oil company may differ significantly from actual behaviour, for strategic or practical reasons. In chapter 2 we categorise different responses on a continuum from reactive to proactive strategic responses on the basis of indicators linked to actual rather than rhetorical behaviour.

We explore three possible reasons for why the climate strategies of major oil companies vary along this continuum. First, the sources of corporate climate strategies may be due to factors linked to the companies themselves. Second, climate strategies may have been caused by the political context of the companies' home-base countries. Third, since the companies under scrutiny are multinational, change in strategies may be the result of changes in the international institutional context in which the companies operate.

Drawing on three bodies of thought – business environmental

management perspectives, theories of domestic politics, and regime theory – we develop a multi-level approach based on three models that may account for differences and change in climate strategies. The first model – *the Corporate Actor model* – simply states that differences in climate strategies are due to differences in company-specific factors such as core business areas, resource reserves, environmental reputation and learning capacity. The second model – *the Domestic Politics model* – postulates that this is not necessarily so, and instead emphasises social demand for environmental quality, governmental supply of climate policy, and political institutions governing company–state relationships in the companies' home-base countries. The third model – *the International Regime model* – takes us from domestic to international politics. Climate change is a global problem largely caused by global target groups. Accordingly, this model takes us from the study of single corporations within single home-base countries to a scrutiny of corporate alliances across states and how they relate to international regimes. Thus, while the CA and DP models to a larger extent are directed towards the analysis of synchronic differences between the companies, the IR model is particularly designed to understand changes in corporate strategies over time.

The term 'model' is not used in a formal sense, but rather to indicate a particular lens through which a simplified picture of the real world is viewed. The three models are all based on the assumption that companies are rational in the sense that their behaviour can be understood as intentional and purposeful action related to survival and profits. The models will be further specified in chapter 2.

The CA, DP and IR models can be tested empirically by means of pattern matching. The strategy is to formulate propositions derived from the models and to compare them with observed changes and differences in corporate strategies (Yin, 1989). The better the match between proposed and observed strategies, the more confidence we will have in the model's ability to explain the phenomena it is designed to explain. However, the length of the causal chain is likely to increase the further we get from company-specific factors. Moreover, we will have to deal with causal complexity (Ragin, 1987). Causal complexity refers to situations where the direction of influence between two variables depends on the value of a third variable. Thus, in situations characterised by causal complexity, the direction of influence is difficult to determine. In these cases, the propositions can be evaluated in terms of explanation-building (Yin, 1989). Explanation-building is based on narratives, and the final explanation may not have been fully stipulated at the outset of the study.

Data collection has been based on multiple sources. Analysis of corporate annual reports and position papers, governmental White Papers and secondary studies has been important. However, interviews with stakeholders in Europe and the US constitute the principle sources of information (see Appendix). We have chosen the following interview strategy: first, we have interviewed representatives of the most important actors, including ENGOs, the oil companies and their branch organisations, and public officials dealing with the oil industry. Second, we have conducted parallel interviews with different representatives of the same oil companies in Europe and the US. For example, we have interviewed representatives for ExxonMobil and Shell in the US as well as in Europe. Since Shell and ExxonMobil are squeezed between two different climate-political contexts in Europe and the US, this strategy has proved very useful in better understanding how the companies actually deal with this situation.

Delimitations

We have chosen to focus on three oil companies: ExxonMobil, Shell and Statoil. These have been selected for a number of reasons. First, and most importantly, Shell, ExxonMobil and Statoil differ significantly with regard to our dependent variable, i.e. corporate climate strategies. Three oil companies that have adopted three different climate strategies allow for a more fine-tuned understanding of the driving forces behind corporate climate strategy choice. Second, these companies are tied to different political contexts: the US, the EU/Netherlands and Norway/Scandinavia. This fact provides us with an opportunity to explore systematically the influence of political context in the companies' home-base countries. Finally, the inclusion of one, until recently, fully state-owned oil company – Statoil – gives us an opportunity to explore whether state-owned companies are driven by other forces in their climate strategies than private companies. For example, one would expect that the link between the climate policy of Norway and Statoil would be particularly strong.

Together with BP, ExxonMobil and Shell represent the three biggest privately owned oil companies in the world and are now often referred to as the three 'super-majors', or 'the three sisters'. The main reasons for excluding BP in this study are twofold. First, BP has adopted a climate strategy quite similar to Shell's. Second, the climate strategy of BP has been studied elsewhere (Rowlands, 2000). The extent to which ExxonMobil is representative of US-based oil companies will be commented upon in chapter 7.

Outline of this book

The composition of this book follows the research questions and the research strategy outlined above. In chapter 2, we present our analytical framework for comparative analysis of the oil companies. We believe that this framework can be useful for analysing other issue areas in which large corporations play an important role. The third chapter seeks to compare and categorise the climate strategies of ExxonMobil, Shell and Statoil. A quick glance at the home pages of these companies reveals significant differences with regard to how they communicate their climate strategies to the outside world. Chapter 3, however, aims to take a critical look at how different these strategies actually are. The aim of chapters 4–6 is to explain corporate strategies. Chapter 4 explores the explanatory power of the Corporate Actor model. Here, we look into factors such as environmental risk, environmental reputation and learning capacity. The domestic political context in the companies' home-base countries is on the agenda in chapter 5. This chapter focuses on the links between Norway and Statoil, the Netherlands and Shell as well as the US and ExxonMobil. In chapter 6, changes in corporate strategies over time are explored by analysing the influence relationship between the international climate regime and corporations. In chapter 7, we conclude the analysis by summarising the empirical findings, reflecting upon the strengths and weaknesses of the research strategy, as well as by drawing some analytical and policy-relevant lessons.

Notes

1 In this study, specific market conditions in the US are seen as the major reason why the American car industry has opposed Intergovernmental Panel on Climate Change (IPCC) climate science and the Kyoto Protocol (see Chapter 7).
2 Skjærseth's (2000) study of North Sea cooperation explores the impact of international and domestic institutions on states, sub national target groups and environmental non-governmental organisations (ENGOs) by relaxing the assumptions of unity and rationality commonly applied in regime analyses.
3 See Paterson (1996) for a study of climate change within the perspective of the grand theoretical international relation debates.
4 For example, Arts (2000) lists a number of studies making this claim.

2
Analytical framework

This chapter outlines the analytical framework of our empirical analysis. Our point of departure is to identify the sources of corporate strategy choice: what factors determine the strategies chosen by the oil industry to meet climate-change challenges? We explore the impact of three main groups of factors, related to: (1) company-specific features; (2) the political context of corporate activity at the domestic level; (3) the international institutional context in which multinational companies operate. Each of these three clusters of factors forms a focus for one of the three 'models' that will be used to shape the analysis. As described in the previous chapter, these are not models in a formal sense, but are rather tools that provide simplified and complementary pictures of the driving forces behind corporate choice at different decision-making levels.

The first model, which we have labelled the Corporate Actor (CA) model, is based on contributions from the business environmental management literature. The CA model focuses on factors that can shape a company's climate strategy, with an emphasis on factors such as environmental risk, environmental reputation and organisational learning capacity.

The second model, referred to as the Domestic Politics (DP) model, is based on the assumption that even multinational companies are heavily influenced by the framework conditions of their home-base countries in which they have their historical roots, have located their headquarters and have their main activities. This model is based on theories of state–society relationships and highlights social demands for environmental quality, governmental supply of environmental policy and the political institutions linking demand and supply. Political institutions shape the channels of interaction between industry, governments and other interested parties.

The last perspective is referred to as the International Regime (IR) model. This model takes us from domestic to international politics, and is based on the assumption that the key sources of corporate strategies are found within the context of international regimes rather than in the political context of the companies' home-base countries. Climate change is a global problem, largely caused by global target groups like the oil industry, dealt with within the framework of international institutions. This model is based on international regime approaches and emphasises how international environmental regimes may trigger changes in corporate strategy choice. The main focus of analysis within this perspective, therefore, is changes in corporate strategies over time. This model thus captures the *dynamic* relationship between multinational corporate actors and international regimes.

Before looking at the models in more detail, let us first explore corporate strategies with respect to climate change.

Focus for explanation: corporate climate strategies

To explain why corporations choose different climate strategies, we first have to distinguish between different strategies. Most of the business environmental management literature operates with a rough distinction between reactive/defensive, proactive/offensive, indifferent, and innovative strategies (see e.g. Steger, 1993). Given a certain environmental risk inherent in a company's activities, a proactive company motivated by profits and survival will exploit market opportunities and support environmental regulation. An innovative company will go further in tapping the market potential by implementing major changes in the production process or by developing new technologies and products. Conversely, a reactive company will deliberately leave market opportunities unexploited and oppose environmental regulation. Indifferent companies will not develop any conscious environmental strategy.

In the context of climate change we need to know precisely what distinguishes the various strategies from one another. A

careful selection of indicators is particularly important in a case
such as this, where the companies under investigation operate
within the same branch and even in the same markets. In this
study, we base our ranking of climate strategies on a continuum
from *reactive* to *proactive* strategic responses, with an emphasis
on what companies actually do rather than on the rhetoric they
use. Companies in between these extremes are referred to as inter-
mediates. This implies that the 'indifferent' category is excluded;
our reasoning is that since oil companies make a living from the
main causes of GHG emissions, climate policy is simply too
important to be ignored. The 'innovation' category is also left
out. Although interesting, it resists empirical analysis in this case,
because technological innovation is so closely related to the operational purpose of the individual companies. Since an innovative
strategy also tends to be proactive, we believe the category of
proactive to be sufficient.

In essence, there are four main ways in which oil companies
can reduce GHG emissions: (1) increasing energy conservation
and efficiency; (2) switching to fuels with lower carbon content;
(3) investing in renewable energy sources; and (4) decarbonising
flue gases through carbon dioxide (CO_2) separation and sequestration. Most large oil companies are engaged in options (1) and
(4), while switching from coal to oil to gas as well as more long-
term investments in renewables represent more significant
changes.

Measuring differences in current climate strategies, however, is
by no means a simple task. First, the public profile of a corporation may diverge significantly from its actual behaviour, for
strategic or practical reasons. Thus, differences in climate strategies may be more visible in their public profiles – the rhetoric they
use – than in their actual operations. Second, the distinction
between rhetoric and actions may be particularly valid with
regard to climate strategies, since climate change is a relatively
new issue area on the political agenda, at least as compared to, for
instance, water and air pollution (which have their roots in the
early 1970s). Changing the behaviour of target groups takes
many years and is normally a matter of incremental change. In the
study of public policy implementation, approximately 10–15
years from the adoption of commitments to evaluation of implementation is recommended (Sabatier, 1986). We therefore have to

rely mainly on a set of 'soft' indicators, which nevertheless can provide us with a good indication of the kind of climate policy futures the three companies are preparing for. Moreover, we will try to move beyond simple rhetoric by giving emphasis to what companies do, particularly in terms of significant investment and divestment decisions with potential long-term implications for the companies' future operations. However, the 'proof of the pudding' lies in actual behavioural change in terms of GHG emissions reduction. Regrettably, reliable and comparative time-series data on CO_2 emissions do not exist for the three companies selected in this study.

We thus consider the following four indicators as the basis for our assessment of the companies' climate strategy choice:

- the corporations' acknowledgement of the problem of a human-induced global climate change;
- their positions on the Kyoto Protocol;
- their GHG emissions targets and measures to achieve those targets;
- the degree of reorientation in their core business areas.

The first indicator is based on the extent to which the companies acknowledge the main conclusions from the IPCC, which, crudely put, states that the problem is real and that there is sufficient scientific evidence to act accordingly. The second indicator is based on explicit announcements. However, public statements made by the companies regarding their stand on the Kyoto Protocol may not accurately reflect their intentions to act. Therefore, we 'control' such announcements by a third indicator pointing to the adoption of (voluntary) GHG emissions reduction targets and measures for their own operations, such as emissions trading. The fourth indicator takes us further towards substance. Here, we make an effort to judge whether a company's response implies a significant reorientation – with long-term implications – in its core business areas. A main focus is on a company's investment or divestment decisions, particularly in terms of decarbonising the company's portfolio. This should give us a picture of the resources with which a company's climate rhetoric is supported. Taken together, these indicators provide us with a sufficiently solid basis to assess and compare the climate strategies of oil companies.

As noted by Rowlands (2000), it is difficult to assess the cause-and-effect relationship between a proactive strategy to climate change and a strategy of decarbonisation of fuels and renewables – it is not at all obvious which is cause and which is effect. If a company divests in coal for competitive reasons and then points to the coal divestment as its climate strategy, we have a classic 'chicken-and-egg' problem. Thus, we will address this problem explicitly according to whether policy or action came first.

The ultimate aim of exploring differences and change in corporate strategies is to understand more about the conditions that promote the shift towards an effective climate policy. As noted in the introduction, large oil companies and the fossil-fuel industry in general represent crucial target groups for a viable climate policy. A proactive strategy among key target groups is a necessary but not sufficient condition for 'environmental problem solving' within the context of climate change. The oil industry can influence the effectiveness of climate policy in at least two ways. First, oil companies can do so by the extent to which they cut their own GHG emissions from upstream activities and improve the quality of products from downstream activities; in contrast to states, a multinational company can require its branch offices around the world to comply with its climate policy. Second, oil companies can achieve this by the way they actively support or obstruct the development of climate policy at national and international levels; companies can lobby against decision-making or lead the way by setting an example for other corporations as well as governments.

Explanatory perspectives

There are essentially two main views on the extent to which large multinational companies are controllable or not within the present state system. On the one hand, some argue that such companies outmatch states in terms of resources and power, consequently operating increasingly beyond the control of nation states. There has been a concern that multinational capital and the growth of international economic institutions, such as the World Trade Organisation (WTO), have circumvented constraints from governments and social movements at the national level (Levy and Egan, 1998). Moreover, there appears to be a mismatch between global corporate activity and democratic governments: while a national government cannot require a part of a company operating in another country to comply with its environmental policy, a multinational company *can* require its parts located abroad to meet corporate standards around the world. In principle, this mismatch in regulatory competence can be dealt with through international environmental institutions. However, while such institutions may have an adequate geographical scope, non-state actors are not formal parties to international agreements and thus not directly committed to international obligations.

On the other hand, others dispute the validity of the claim that international economic integration or globalisation has produced the 'global corporation', which owes allegiance to no state. Rather, they posit that multinational corporations operate within enduring political structures that continue to account for striking differences between them (Doremus et al., 1998; Pauly and Reich, 1997). Multinational corporations are not only under the control of all the states in which they operate, they are also largely controlled by their home-base countries in which they have their historical roots, headquarters location, and frequently main operations. A government has the authority to set standards and enforce regulations for all entities within its borders, while a company cannot set standards and require other companies to comply with them.

These opposing claims have some important implications for environmental governance. The first claim implies that global corporations are virtually out of regulatory reach in the present world system – characterised as it is by the absence of a central global authority. The latter claim is more optimistic in terms of governance: global corporations can be controlled within the existing frameworks of the nation state and by international regimes.

The three models presented below will shed light on these different claims. If large multinational oil companies tend to operate beyond national control, we would expect that the CA model has high explanatory power. The CA model suggests that differences between the companies themselves are more important than differences in political context for explaining the strategies they choose. The two other models are linked to the second claim. The DP model indicates that global corporations are controlled by their home-base countries. The IR model suggests that multinational companies are influenced by the rules, norms and procedures of international regimes, and that these regimes can affect corporations directly by shaping collective corporate expectations, or indirectly by harmonising regulations in member countries.

The Corporate Actor model

As the demand for industry to give a higher priority to environmental issues has increased, a separate business environmental management literature has been developed. A number of models and approaches have been produced to explore and explain the environmental strategies and performance of corporate actors.[1] This body of literature is relatively new and is still to a large degree in the conceptual and exploratory phases. However, three main sources of influence on corporate strategy can be identified: (1) factors linked to the political and legal context within which the companies operate; (2) factors linked to the business context of the companies; and (3) company-specific features. Factors linked to political and legal contexts will be included and specified in the DP model, discussed below. In this section we will focus on the latter two sources of influence, looking specifically at three main factors relevant for a company's choice of climate strategy:

- the environmental risk associated with the company's activities;
- its environmental reputation;
- its capacity for organisational learning.

As pointed out by Steger (1993), the main concern of businesses is survival or profit maximisation. It is thus reasonable to assume that a consciously developed strategy to improve the corporation's environmental performance enters into the equation only when poor environmental performance threatens to undermine long-term survival. Steger thus views a company's environmental strategy as being determined by the level of *environmental risk* inherent in the company's activities. Environmental risk is in turn assumed to interact with market opportunities provided through environmental protection. In our framework, however, provision of market opportunities by means of renewable energy policy, for example, is defined as a part of the political context in which the companies operate rather than as part of the companies themselves.

As are other industrial sectors, the oil industry is presently targeted by environmental regulations covering air, water and soil in every link of the production chain, from exploration to retail distribution. Regulations of product quality or emissions to air or water seldom represent a threat to the survival of oil companies, although the stringency of regulations varies from country to country (Estrada et al., 1997). In contrast to 'traditional' environmental regulation, the risk of regulation faced by oil companies in the field of climate change may be more lenient in the short term, but more severe in the long term. In addition to regulations directed at GHG emissions from oil industry activities, stringent international regulations may also affect the volume of production, since the oil industry earns its livelihood from the main sources of GHG emissions. Following the same logic, we have to search for nuances in the companies' fossil-fuel portfolio in order to understand differences in climate strategies.

Most multinational oil companies produce oil, gas and coal. Coal is the most carbon intensive, followed by oil and gas respectively. According to Rowlands (2000), it is reasonable to assume that the more carbon intensive the fossil-fuel portfolio of the oil companies, the higher the risk of the companies being targeted by stringent regulation and the more likely they are to resist such policies. The *relative* importance of coal, oil and gas will thus determine climate strategies: companies with more emphasis on coal and oil are more likely to adopt a reactive climate strategy than companies with a larger relative emphasis on gas. Carbon intensity is explored in terms of core business areas, exploration and production volume as well as resource reserves.

A company's perceptions of risk is also linked to another key factor: its *environmental reputation*, including its experience with public exposure and criticism in relation to environmental incidents. Such criticism may damage the brand name of a company. Environmental reputation may affect climate strategies directly and indirectly. A direct causal pathway may be discerned in the sense that companies with experience of strong public criticism stemming from severe incidents, such as the *Exxon Valdez* spill (see chapter 4), will seek to avoid negative public scrutiny by adopting a proactive climate strategy. In this way, a company may respond effectively to an enhanced public concern for climate change. A poor environmental reputation may also affect companies indirectly, by initiating a reorganisation process aimed at streamlining the implementation of environmental standards within the company in order to prevent parts of the company from damaging the reputation of the whole. In turn, this may also stimulate a proactive climate strategy (see below). On this basis, we assume that a negative environmental reputation induces companies to choose a proactive rather than a reactive climate strategy.

While these factors are company-specific, they represent external sources of influence on a company's environmental strategy choice. Another set of factors that may have an impact on strategy choice and corporate environmental performance stems from internal sources. Here we focus on a company's capacity for *organisational learning*. Organisational learning basically concerns two main dimensions (see also Post and Altman, 1992; Neale, 1997): First, a company's capacity to learn depends on its openness towards its external environment; that is, the degree to which it exposes itself to the outside world and its capacity to capture signals of trends in areas of relevance to its business. Thus, the extent to which a company has institutionalised a systematic monitoring of future trends is one important determinant of the company's learning capacity. Second, a company's capacity to learn also depends heavily upon its capacity to make use of – internalise – the knowledge generated through monitoring mechanisms. This dimension of learning thus concerns the extent to which the organisational structure of the company facilitates effective intra-organisational communication and coordination.

The first dimension of organisational learning – the institution of monitoring systems in the organisation – may be decisive for the future of energy supplies companies prepare for. For instance, why have some major oil companies redefined themselves towards *energy* companies, with a stronger focus on non-fossil-fuel energy sources, while other companies continue to focus exclusively on fossil fuels? We assume that this may have something to do with differences in the companies' emphasis on and interpretation of the 'shadow of the future' – i.e. the companies' capacity and willingness to understand a changing world.

Since fossil fuels represent a non-renewable energy source, most multinational oil companies monitor future reserves and markets. Companies may nevertheless differ along two dimensions. First, they may differ in the extent to which and how monitoring of future trends is institutionalised within the organisation and used systematically as a decision premise within the organisation. Such systems are likely to be directed at understanding future markets, consumer preferences and opportunities, and risks arising from relevant political contexts. Second, even though companies have institutionalised such systems to the same extent, they do not necessarily have the same vision of the future. What they see will depend on previous experience with related changes. For instance, if two oil companies both see a window of opportunity for solar energy, they may still respond differently depending on whether they have positive, negative or no past experience with such technology. What they see will also depend on where they look: whether at their own history or at features characterising the context in which they operate.

The extent to which the companies are capable of making use of the information generated through monitoring systems is to a large extent linked to the organisational structure of the company, with a rough distinction running between centralised and decentralised companies. It could be argued that a centralised company is better equipped for internal communication and coordination, and thus has a larger capacity to make internal use of information generated through monitoring. A decentralised company, on the other hand, would be less capable of communicating trend shifts from one part of the organisation to another, and would thus also be less capable of internalising this kind of information.

These two dimensions of organisational learning are thus both necessary for a company's learning capacity: without a certain degree of openness towards the external environment and a systematic approach to monitoring future trends, a company is incapable of identifying relevant trend changes when they occur. Similarly, without effective channels of internal communication and coordination, the company is incapable of making use of the knowledge generated. In addition, as discussed above, a company's interpretation of future trends also depends upon its history and past experience.

This indicates that the relationship between learning capacity and corporate strategies is characterised by causal complexity and is hence difficult to determine (Ragin, 1987). Causal complexity refers to a situation where it is the combination of conditions that produces change – which is different from saying that each variable in itself produces a change in another variable. When the relationship between a set of variables is characterised by causal complexity, the *direction* of influence, which in our case would be whether one variable leads to a proactive or a reactive strategy, depends on the value of another factor or variable. This implies, first, that a company has a high learning capacity only to the extent that it both monitors future trends and has an organisational structure equipped for communicating and coordinating the insights from its monitoring activities. In addition, how information generated through monitoring activities is interpreted may depend upon the company's history and previous experiences. The presence of causal complexity implies that similarities in learning capacity produce similarities in climate-strategy choice only in combination with similarities in other factors. At the most general level, however, we nevertheless expect that a company that anticipates a significant role and demand for renewable energy sources in the future and has the organisational capacity to internalise this vision is more likely to adopt a proactive climate strategy.

In general, we assume that similarities in company-specific features will lead to similarities in the companies' responses to climate change. Likewise, we assume that variation in company-specific features will lead to variation in the companies' responses to climate change. More specifically, we propose that a low level of environmental risk, negative public scrutiny, and high organisational learning capacity (conditioned by other factors) will lead to a proactive strategy on climate change.

Moderating factors   In addition to the three factors chosen here, scholars in business environmental management have suggested a set of company-specific factors that may determine environmental strategies, including leadership, capital availability, human resource availability, corporate tradition and ownership (see inter alia, Ghobadian et al. 1998). With respect to leadership, there are a number of references to the unique role of Sir John Browne in directing BP towards a more proactive stance on climate change (see e.g. Rowlands, 2000). Differences in leadership may, however, be prohibitively difficult to assess empirically in relation to climate strategies in a comparative perspective. The same is true when it comes to capital availability. Kolk and Levy (2001) argue that low profitability may lead to a reorientation towards renewables. On the other hand, low profitability may also lead to caution concerning new and more risky investments. Human resource availability can be expected to be roughly equal for global companies operating in the same global market. However, there may be differences in in-house scientific and technological expertise that may influence the perception of causes as well as solutions to environmental problems characterised by scientific uncertainty.

Finally, an important organisational dimension with a potential impact on environmental strategy choice is the *ownership structure* of the corporation. First, there is a major distinction between state and private ownership. Shell and ExxonMobil are private companies, while Statoil was, until recently, fully owned by the Norwegian state. National oil companies may be less accountable to the capital market, and even when they are exposed to competition, they often operate in a privileged position with close consultative relationships with their government owners, who also regulate the industry (Noreng, 1996). It is difficult, however, to establish the weight of this factor and even the direction of its impact, but it seems reasonable to assume that government-owned companies are likely to adopt a climate strategy in accordance with the position of their government owners. Second, ownership may have an impact on strategy choice in the sense that shareholders in private companies may pressure the corporate leaders to adopt a more proactive strategy. Third, differences in shareholder patterns between Europe and the US may have implications for the extent to which corporate leaders adopt a long-term perspective in their financial decisions, or whether they merely focus on short-term shareholder returns. Such differences may have relevance for the companies' climate strategies (Pauly and Reich, 1997). This is extremely difficult to explore empirically, however, because the causal chains are long and complex.

The Domestic Politics model

The DP model is a well-established approach within political science (Underdal and Hanf, 2000). It suggests that key sources of state behaviour can be found at the domestic political level rather than in the international society: differences in state responses to common problems may be traced back to the state, or government itself, the society, or the relationship between state and the society. We have used this model to shed light on how multinational corporations adopt climate strategies. Accordingly, we assume that corporations are affected by a social demand for environmental protection, governmental supply of climate policies and the political institutions linking supply and demand. Notice that the DP model was originally developed to understand political decision-making rather than corporate decision-making. This difference in what the models are designed to represent has particular consequences for the social demand dimension (see below).

Multinational oil companies are potentially affected by social demands and policies in all countries where they operate. However, some researchers have argued that the 'nationality' of private multinational companies is of particular importance for their attitudes and culture (Gleckman, 1995; Rowlands, 2000). The strongest influence is likely to be found in the companies' home-base countries, where they have their historical roots, have located their headquarters and have concentrated most of their activities. This observation is supported by a survey of multinationals revealing that the most important motivating factor for establishing corporate-wide environmental management systems is the environmental policy of the companies' home-base countries (Gleckman, 1995). Thus, it seems reasonable to assume that long-standing national ties affect the way in which companies approach new problems such as climate change.

Social demand   The key mechanism whereby social demands can affect the actions of governments is the voting power of electorates. In this analysis we transform social demand from being an analytical tool for understanding governmental behaviour to one for understanding corporate behaviour. The key mechanism whereby social demands are assumed to affect corporate strategies is thus consumer behaviour rather than voting power.

A social demand for environmental protection affects corporations engaged in activities associated with environmental risk (Rondinelli and Berry, 2000). Public values and attitudes as well as organised social interests, such as environmental groups, control a powerful tool for inducing specific modes of corporate behaviour: consumer behaviour. 'Green' consumerism has become a significant force in Organisation for Economic Cooperation and Development (OECD) countries, and this phenomenon has the capacity to stimulate and weaken product markets. There are at least two mechanisms through which social demands may affect corporate strategy choice on environmental issues. First, consumer campaigns initiated by the green movement can damage companies' reputation and affect their market share. Second, in their choice of environmental strategy, companies may be responding to 'green' consumers' willingness to pay a higher price for clean products, such as clean energy.[2] While the latter mechanism provides companies with business *opportunities* like new markets in renewable energy, the former exposes companies to *pressure*. A strong social demand for climate policy is thus likely to trigger opportunities and pressures simultaneously. Whenever national imprints overlap with market exposure in terms of pressure and opportunities, we suspect that social demands in the companies' home-base countries will influence their environmental strategies.

A social demand for a viable climate policy is likely to affect corporate strategies only marginally if changes in values and attitudes are perceived as expressions of short-term fluctuations only. According to Inglehart (1971), support for different social movements, including the environmental movement, represents a political expression of post-materialistic values that will strengthen their position on the political agenda as the new generations become older. We would thus expect that the strength of a social demand for environmental protection would increase gradually in the form of 'new' values and attitudes. In contrast, Downs (1972) proposes that environmental issues, like other political issues, will follow an 'issue-attention cycle'. Environmental issues will fade from the interest of the public over time and be replaced by new issues, regardless of whether problems actually have been solved. The point here is that the relevance of these explanations can vary between countries, providing corporations with different signals with regard to what can be expected in the future concerning people's willingness to pay higher prices for clean energy. For example, fluctuations in environmental attitudes in Norway and the Netherlands – the home-base countries of Statoil and Shell – have been interpreted in line with Downs and Inglehart respectively (Weale, 1992; Aardal and Valen, 1995).

The fact that the key mechanism through which companies may be affected by social demands is consumer rather than voting behaviour implies that there is also an indirect pathway through which social demands can affect target groups – namely, through public policy. Public pressure is in itself a contextual factor shown to be important for explaining outcomes of national environmental policy (Jänicke, 1992, 1997). Since politicians in democratic systems are accountable to their electorates, change and variance in public environmental values and attitudes to climate change may affect governmental climate policy. Thus, societies characterised by a strong social demand for environmental protection are also more likely to enforce stronger environmental regulation. Social demands, therefore, can affect both the strength of climate policy (with an indirect effect on companies' strategy choice) and consumer behaviour (with a direct effect on companies' strategy choice).[3] Companies are thus sensitive to the social context in which they operate for many reasons. On this basis, it is reasonable to assume that a strong social demand for environmental quality will stimulate a proactive strategy. Conversely, a weak social demand for environmental quality is likely to go hand in hand with a reactive strategy.

Governmental supply In democratic systems, social demands represent a significant force in shaping public policy. Public policy, however, is not only driven by social demands. Governments have both their own views and the capacity to act independently. Governmental regulation has been seen as an important factor behind the 'greening' of industry since the UN Conference on Human Development in Stockholm in 1972 (Falkner, 1996).

Governmental supply of an ambitious climate policy is here understood in terms of targets and policy instruments. The world is full of political declarations that are not seriously intended and never realised – a well-known phenomenon from large diplomatic conferences and election campaigns. Even well-specified climate targets, including deadlines and baselines, do not necessarily send any clear signal to target groups unless they are backed up with policy instruments. Climate-policy instruments represent the 'sharp end' of public policy and are particularly important for companies' climate strategies.

Three broad categories of environmental policy instruments have evolved over the past 30 years (OECD, 1999).[4] The first is regulatory instruments – often referred to as 'command and control' – whereby public authorities mandate a certain performance or technology. The second is economic instruments whereby target groups are given financial incentives to reduce environmental damage. Voluntary agreements constitute the third type. These types of policy instruments resemble both the stick and the carrot, as well as agreements at the interface of sticks and carrots: the government may force us, pay us or have us pay, or persuade us to strike a deal in the 'shadow' of hierarchy.

In the 1990s, voluntary agreements between governments and industry received increased attention in climate policy, and more than 350 voluntary programmes have been adopted in 22 OECD countries (IEA, 1997). Voluntary agreements involve commitments by target groups to improve their environmental performance beyond what is strictly legally demanded. There are two main types of agreements: negotiated agreements and public voluntary programmes. Negotiated agreements are binding, highly structured, and developed through bargaining between a public authority and industry. In contrast, public voluntary programmes are optional commitments in which companies are invited to participate (OECD, 1999).

Different policy instruments have different qualities in accordance with different problems, and may be judged according to a number of criteria, such as goal attainability, capacity to stimulate technological innovation, cost-effectiveness and transparency (Skjærseth, 2000). However, the *authoritative force* of policy instruments appears particularly important for companies' climate strategies. Authoritative force represents the degree of constraint the government exercises on the target group, affecting the group's discretionary room for manoeuvre (Vedung, 1997).[5] Policy instruments based on a high degree of authoritative force, like direct regulation or economic instruments, send a clear signal to target groups: the authorities acknowledge the problem and expect companies to change their behaviour accordingly. In contrast, public voluntary programmes are normally used as a first step in the exploration of a new policy area (OECD, 1999). Thus, these programmes tend to be associated with a high level of uncertainty with regard to future regulation.

A viable climate policy based on clear targets and mandatory policy instruments can reduce uncertainty, create regulatory pressure and grant market opportunities for companies. Reduction in *uncertainty* concerning future options is particularly important for the oil industry, which earns its livelihood from non-renewable resources expected to run out sometime in the future. Predictability in regulatory frameworks is important when oil companies make decisions on their own climate targets, abatement measures or investments in renewable energy. Since 1971, Shell has, for example, addressed uncertainty in its strategy formulation through scenario planning (see chapter 4). A proactive response can also be seen as a function of regulatory *pressure* exercised within a political context of the increasing importance of insurance companies, responsibility and liability. The oil industry has experienced a gradual strengthening of environmental policy since the 1950s (Estrada et al, 1997). Today, every link in the oil industry chain – exploration and production, transportation, refining and distribution – is controlled by a series of regulations aimed at preventing water, air and soil pollution. Regulatory pressure creates corporate attention, which represents the first step towards any conscious climate strategy. Pressure further induces unilateral company targets as well as abatement efforts. Neglecting regulation exposes oil companies to economic risks. A proactive strategy may reduce this risk: 'the more the oil industry opts for a 'wait and see' approach, the more it is likely to attract the attention of regulators' (Estrada et al., 1997: 16). Thus, companies sometimes adopt a proactive strategy in order to gain more favourable treatment by regulatory agencies (Rondinelli and Berry, 2000). In addition, climate strategies are likely to be influenced by previous corporate experience with environmental regulation in other issue areas. An ambitious climate policy can also create market *opportunities*. Governmental targets and measures on renewable energy will provide the industry with incentives to focus accordingly. In addition, increasing abatement costs may stimulate development of commercially viable technology on energy efficiency.

The upshot of the above reasoning is the proposition that an ambitious climate policy in terms of targets and policy instruments will stimulate a proactive strategy among oil companies. If high social demands go hand in hand with an ambitious climate policy, a positive *interplay* between demand and supply can be expected. Conversely, low demands and lenient policy will point in the direction of a reactive strategy.

Political institutions linking demand and supply   Western democratic political systems have two main channels for influencing decision-making by linking state and society: the numerical-democratic channel, which includes voters, political parties and parliaments; and the corporative channel, in which non-governmental and governmental decision-makers meet to consult, collaborate and negotiate. The relative importance of each of these channels has traditionally been summed up in the phrase 'votes count, but resources decide' (Rokkan, 1966). Thus, here we concentrate on the presumably more influential and relevant of the two: the corporative channel.

There are different theories of business–state relations. One view is that the state actively serves business interests that are able to act cohesively in the political arena. Another is that the state can maintain neutrality and independence from business interests (Levy and Egan, 1998). This section is based on the latter assumption, but we pragmatically adopt the view that this is essentially an empirical question that may lead to different answers in different cases.

Governmental decision-makers are, at least in theory, left with a choice between stimulating cooperation aimed at consensus-building between industry and governmental decision-makers, and a more conflict-oriented strategy based on imposition. The distinction between political institutions based on consensus or imposition is a fundamental one in the study of comparative environmental politics (Lundqvist, 1980; Jänicke, 1992; Andersen, 1993). Such distinctions are often referred to as significantly different national styles or approaches to regulation. Political institutions determine who is included in the decision-making processes, to what extent and in what way. A consensual approach is characterised by open formal access for affected target groups, such as the oil industry. The aim of this collaborative strategy is to raise environmental awareness and promote social responsibility among companies. A recent trend in environmental policy highlights consensual principles such as 'shared responsibility' between firms and governments. In return, companies expect their interests to be taken into account in the design of relevant policy. Accordingly, we propose that a consensual approach stimulates a proactive corporate strategy. In contrast, a conflict-oriented approach is characterised by limited access to decision-making in climate policy for target groups. The rationale behind this approach is to avoid regulatory capture, i.e. the regulated taking control over the regulator. A conflict-oriented strategy is likely to produce resistance among target groups in the form of a reactive strategy.

Political institutions that stimulate a proactive approach do not necessarily lead to higher environmental effectiveness. With the conflict-oriented strategy, industry will have a limited opportunity to water down regulations, although opposition during implementation can be expected; if target groups have not been invited to have their say, their interests are less likely to be reflected in the goals and composition of policy instruments. Conversely, the collaborative strategy is likely to lead to cooperation during implementation, but often at the expense of more lenient goals and policy instruments. In short, these approaches may lead to different corporate strategies, but a similar outcome in terms of environmental effectiveness.

While the main focus here is how political institutions affect corporate strategies, corporations not only represent a potential target for social demands and governmental policies, but represent in themselves a social interest group with a potential to influence governmental policies. This is not least reflected in the emphasis given to the DP model in the emerging literature on the role of multinational actors in policy-making (see, for instance, Risse-Kappen, 1995). Corporations themselves may have the potential to influence governmental policies in specific issue areas. There is, however, no necessary relationship between formal access and influence. First, informal participation in the form of lobbying is widespread throughout the industrial world. Second, the insider/outsider model suggests that some actors or alliances may enjoy a privileged status and represent the 'core' while others are more peripheral (Maloney et al., 1994). Still, formal access tends to stimulate a constructive, cooperative climate and increase probabilities of influence, since it does not necessarily exclude other forms of informal participation.

Companies may have good opportunities to affect the ambitiousness of climate policy particularly in their home-base countries, but such influence is probably less significant *in itself* as a determinant of corporate strategies. In the case of international regimes and global problems, however, corporate influence on the climate regime may in itself prove crucial for their collective climate strategies. This mechanism will be explored in the next section.

On the basis of this discussion of the DP model, we may generally assume that if domestic political context factors differ and are decisive for corporate strategy choice, corporations will tend to choose different, rather than similar, strategies towards common problems. More specifically, we explore the proposition that strong societal demands for climate policy, governmental supply of an ambitious policy, and a consensus-oriented approach to regulation will promote a proactive strategy among multinational oil companies.

The International Regime model

The IR model takes us from domestic to international politics. According to this model, the key sources of corporate strategies are found within the context of international regimes rather than in the domestic political context of the companies' home-base countries. Climate change is a global problem, largely caused by global target groups such as the oil industry, dealt with within the framework of international institutions. We thus have to move beyond the study of single companies within single states in order to gain an understanding of *changes* in corporate climate strategies. Accordingly, this model is concerned with corporate alliances across states and how such alliances relate to international regimes. The DP model is more static and directed towards explaining differences in strategies rather than changes.

According to the regime perspective, corporate climate strategies are likely to be formed by the influence of industry on regimes and the influence of regimes on industry. This argument is straightforward: if industry determines joint international commitments, the regime is in turn unlikely to affect industry strategies. Conversely, if industry exercises little influence on joint international commitments, the regime has in turn a potential to affect industry strategies. The IR model presented here is thus concerned with the *dynamic relationship* between corporate strategies and international institutional development. By dynamic, we mean a relationship in which institutions can affect the strategies of corporate actors, which in turn can affect institutions, or vice versa. This approach thus allows for a more systematic focus on the conditions triggering changes in corporate strategies.

In contrast to the national level, no central governing body exists at the international level that has the authority to define stakeholder involvement, solve disputes among affected actors or enforce regulations. The international system has been characterised as an anarchy based on a self-help system among *states* (Waltz, 1979). States can modify the self-help element of international relations by transferring authority to international regimes, such as the climate regime. International regimes can be defined as: 'social institutions composed of agreed-upon principles, norms, rules, and decision-making procedures that govern interactions of actors in specific issue areas' (Young and Osherenko, 1993: 1).

International regimes provide industry with both constraints and opportunities. On the one hand, corporations have limited influence over the development of international regimes because states, not companies, are parties to regimes. Since international environmental regimes tend to 'mature' over time by developing more stringent joint international commitments, regimes represent a serious challenge for reactive companies (Miles et al., 2001). A strong global regime carries the potential of affecting multinational companies all over the world, in sharp contrast to the policy of one single state. On the other hand, international institutions based on *agreement* may simply codify the behaviour of the most 'reluctant actor' – a mechanism known as the 'law of the least ambitious program' (Underdal, 1980; Sand, 1991). If a reactive industry determines the position of the 'least ambitious' state in international negotiations, and that state determines joint international commitments, the regime is unlikely to stimulate proactive corporate strategies. This mechanism provides (a reactive) industry with the opportunity of exercising influence far beyond its home country. Alternatively, industry influence can affect the inclusiveness of the regime: reluctant states can choose the exit option, or the regime can allow for various 'fast-track' options allowing different obligations for a subset of the parties (Sand, 1991).

If the regime is based on qualified majority, single reluctant states can be outvoted and the influence of a reactive industry on the regime tends to decrease. In essence, the perceived influence of industry appears crucial for choice of climate strategy. In situations where a reactive industry has weak influence and the regime 'matures', we can expect a change towards a proactive strategy. In this case, there will be more to lose than to gain from a persistent reactive strategy. Conversely, if a reactive industry exercises strong influence and largely controls the development of the regime, a persistent reactive strategy can be expected. However, we cannot exclude the possibility that a company or branch may move towards a proactive strategy *prior* to any regime development. In this case, however, corporate strategies can influence regime development and not the other way around. One example here is the change in the strategies of the chemical industry, which contributed to a breakthrough in the ozone regime (Skjærseth, 1992). Such a development would not have been predicted merely by the IR model, but could be explained by additional factors linked to the DP or CA models.

In situations where industry exerts a weak influence on the regime, the regime can in turn also explain differences in corporate strategies to the extent that it generates different rules, norms and principles for different (groups of) actors. This factor is primarily linked to the inclusiveness of the regime: are all relevant/pivotal state actors included as members? As governments, regimes can provide companies with opportunities or pressure, though through different pathways. Moreover, international regimes can influence industry by producing knowledge on the causes and consequences of the problem at hand.

From the perspective that large multinational corporations actually cause a significant number of the problems international regimes have been established to solve, the state-centric approach of regime theory is a major shortcoming. Although ENGOs have received increased scholarly attention, systematic studies of multi-national corporate actors are almost non-existent in the regime effectiveness literature (see chapter 1). Thus, the next two sections will have a more exploratory status than the former two. Here, we will make an effort to include corporate actors in the study of international regimes by focusing on two key questions: to what extent and how can oil companies affect the climate regime? And how can the climate regime affect the strategies of the oil companies?

Corporate influence on international regimes   Studies of international regimes have evolved rapidly, from focusing on regime creation to focusing on regime effectiveness. Different approaches within regime effectiveness theory are unified by a common assumption that international institutionalisation within specific issue areas can affect the will and ability of states to come to grips with common challenges. The concept of effectiveness has been defined in a number of ways, but all definitions direct our attention to the *consequences* of regimes (Young 1994; Underdal, 1992, 2001).[6] A starting point in the study of how corporate actors can affect regime commitments is to turn the conventional chain of regime consequences upside down: our point of departure is the strategies of target groups rather than the consequences flowing from international regimes. This means that corporate actors can influence regimes through two main channels: the domestic and the international regime. Domestically, corporate actors can affect (1) the formation of national positions in international negotiations, and (2) the domestic implementation of joint commitments. In addition, corporate actors can affect (3) international cooperation directly through their presence (as observers or lobbyists) at international negotiations. In this section, we focus on the extent to which and how corporations can affect the stringency and geographical scope of joint climate commitments. The focus on joint commitments rather than their domestic implementation is necessitated by the short lifetime of the climate regime.

Notice that the main distinction between (1) and corporate channels at the domestic level lies in a shift in focus from national policy to the formation of national positions in international negotiations. In environmental policy, there is a probable but not necessary relationship between national policy and international positions. Norway has, for example, been more ambitious in its environmental foreign policy abroad than at home (Skjærseth and Rosendal, 1995).

Determining the influence of a specific branch of corporate actors is extremely complicated. First, industry affects national positions and joint commitments both indirectly and directly (Newell, 2000). Indirect *structural* influence is related to states' dependence on industry: industry is important for economic growth, employment and technological innovation – particularly in the energy sector, which tends to be viewed as a strategic state objective. This structural dependency provides industry with privileged and informal access to decision-making, which is difficult to observe directly. Direct *instrumental* influence is based on huge in-house human, financial and technological resources, which are deployed to persuade decision-makers through PR firms, disputing climate science and developing economic models showing the high public costs of regulating GHG emissions. Moreover, large corporations have the ability to organise at all levels of society. Global companies such as Shell and ExxonMobil are particularly well suited to match global environmental regimes, and global presence is secured in business organisations such as the International Chamber of Commerce (ICC).

Second, the notion of industry influence is difficult to pinpoint precisely because the concept is closely related to power, and because power and influence have mainly been related to states. Power has in state-centric terms been related to capabilities, while influence is seen as a relationship between actors, which can modify behaviour (Cox and Jacobson, 1973). In the same way, we may say that industry possesses structural and instrumental capabilities. As pointed out by Betsill and Corell, however, the important question is how capabilities are translated into influence (Betsill and Corell, 2001: 74). These researchers' answer is to relate influence to persuasion when one actor intentionally transmits information to another that alters the latter's actions from what would have occurred without that information.

Third, it is difficult to separate the influence of one particular corporation or industry from that of others since companies tend to coordinate their positions within industry organisations. Thus, we take a broader view on industry in general, particularly the fossil-fuel industry, in which the major oil companies have played an important if not dominating role.

At least three conditions related to the qualities of the target groups, institutions and other actors are important for understanding the influence of industry on joint regime commitments. These conditions are thus important for how industry perceives its influence. First, we assume that the influence of target groups tends to increase the more cohesive their strategies are. Political and financial resources, activities, and the strategies employed by target groups are likely to enhance influence if target groups stand united in their support for or opposition to specific policies. Target groups participate in international environmental politics in different ways: they organise themselves at all levels of society to affect policy whenever their interests are threatened; they lobby decision-makers when positions in international cooperative efforts are shaped; and they participate as observers in international negotiations (Newell, 2000).

Second, there is reason to assume that influence depends upon the *access* industry has to decision-making processes as well as the *decision-making procedures* applied. Since corporate actors can influence international regimes through the domestic as well as the international regime channel, access to decision-making processes relevant to the development of national positions in both international negotiations and the regime is important. Even though corporations do not participate in international regimes, their access as observers to preparatory sessions can vary significantly. We assume that influence will increase as industry has more access to decision-making processes. Decision-making procedures applied in international environmental cooperation vary as well. Unanimity is most demanding, requiring the positive approval of all parties. Under the condition of unanimity, reactive industry influence within one single state can block the efforts of all others. The requirement of consensus is less demanding in that it merely requires the absence of objections. Consensus is often used in combination with various 'fast-track' options, such as the principle of differential obligations and regionalisation of the cooperation (Sand, 1991). These opportunities limit the influence of a reactive industry to block the efforts of merely a subset of the parties. In the case of a qualified majority decision binding on the minority, reactive industry influence through domestic channels will be further limited. In the EU, for example, a minority block of EU member industries can be forced – via reluctant national authorities – to implement a specific directive or regulation. If an industry fails to comply, it may be brought before a national court, which is required to interpret national laws in line with EU obligations. In addition, since 1993 the European Court of Justice has been empowered to impose fines on states that have failed to comply with previous rulings of the court.

Third, influence is likely to depend on the strength of counter-balancing forces, such as ENGOs. The environmental movement tends to represent a significant counterbalancing force to target groups. In essence, the more ENGO resistance to business strategies, the less industry influence can be expected. As in the case of industry, ENGO resistance can be exerted though domestic as well as international channels.

Regime influence on corporate strategies   In this section, we shift the focus to how international regimes can affect corporate strategies, with a particular focus on changes in strategy over time. The dynamic twoway relationship between international institutions and corporate actors represents a fascinating aspect of climate policy and contains the seeds of change towards a more effective climate policy in the future. The study of institutional dynamics within regime theory has generally been related to the grand questions of 'The rise and fall of international regimes' (Young, 1989). Institutionalised cooperation may lead to path-dependent processes where earlier events affect subsequent ones. Our focus will be narrower, but feed into the topics indicated above.

The question in this section is whether and how the institution governing climate change has constrained or promoted subsequent actions among corporate actors due to its initial qualities. As noted, international regimes tend to 'mature' over time towards more stringent joint commitments. This observation is in line with the notion that institutionalised cooperation gathers momentum through a 'snowball' effect generating positive feedback and facilitating further steps (Andresen et al., 1996). According to Young, international regimes evolve continuously in response to their own inner dynamics (Young, 1989: 95). Levy has labelled such dynamics of international environmental institutions as 'tote-board diplomacy' (Levy, 1993). According to this perspective, we can expect that, over time, joint commitments will become more stringent and governments as well as corporations more ambitious in implementing them. It is reasonable to assume that this process becomes more likely when the initial institutional arrangements have a narrow scope, include lenient commitments and possess institutional feedback mechanisms that encourage dynamic development.

Even though most regimes tend to 'mature', others may decline. This possibility is reflected in the economists' 'law of diminishing returns'. According to this perspective, the first steps are likely to be the 'easy ones' in which marginal benefits clearly exceed marginal costs. Then when attempts are made to extend the scope of the institution or to tighten up joint commitments, marginal abatement costs will tend to increase, and benefits in the form of environmental quality will tend to decrease. For example, according to a rule of thumb applied by industry, costs will remain constant for each 50 per cent cut in discharges. This perspective would suggest that it will become increasingly difficult to step up joint commitments, and governments as well as corporations will become more reluctant in implementing them over time. The conditions triggering this development are more likely to appear when original institutional arrangements have a wide scope and include stringent commitments as well as dynamic qualities.

How can international regimes induce change in corporate strategies? We believe that the core mechanism may lie in the combination of the 'snowball' effect at the regime level and the lack of industry influence in international regimes. On the one hand, corporations increasingly participate as observers of international regimes. The UNFCCC places strong emphasis on participation by non-state actors, and the Global Climate Coalition (GCC) alone came with a delegation of 50 members to Kyoto (Raustiala, 2001). On the other hand, the principle of sovereignty provides states with significantly more rights than large corporations. States have the right to refrain from international agreements that they have not given their consent to. Sometimes states also have the power of veto and the right to block the efforts of other states. Corporations do not have these rights, even though a company like ExxonMobil is much larger in terms of economic resources than many parties to the UNFCCC. ExxonMobil dislikes the international climate process and finds the international climate regime unrepresentative.[7]

We analyse the climate regime (the UNFCCC and the Kyoto Protocol) and the climate policy of the EU. The EU can be analysed both as an actor in the international climate regime and as a regional subregime (Skjærseth and Wettestad, 2002). We explore three causal pathways through which conditions linked to the international regime may have affected corporate strategy choice: knowledge, pressure and opportunity. These are compatible with three well-known causal mechanisms applied within regime theory: knowledge, power and interests (Young and Osherenko, 1993). Note that some of these regime qualities may serve the same functions as domestic policies. The main difference is that international regimes are able to carry out these functions on a wider scale than individual countries. In other words, a global regime has the capacity to affect global multinational companies in all countries in which they operate.

First, there is a scientific/technical *knowledge-based* pathway, which affects the extent to which corporations accept a common understanding of the problem at hand. The scientific uncertainty argument has repeatedly been used by the corporate fossil-fuel lobby to oppose any attempts to adopt a viable climate policy. We assume that differences along this dimension may be due to either differences in corporate access to the IPCC process, or differences between the corporations in their receptiveness to the information provided. Significant changes in the scientific knowledge base may also bring about corresponding changes in corporate strategies.

Second, international regimes may exert regulatory *pressure* directly on companies or indirectly by strengthening national climate policy. The general assumption is that 'strong' regimes will promote a stringent climate policy and proactive companies by shaping mutual expectations about the need for future regulation. In essence, strong regimes send a clear signal to target groups. Regime strength can be seen as a function of: (1) the authoritative force of commitments; (2) ambition and specificity of commitments; and (3) verification and enforcement systems.

Third, international regimes can also grant *opportunities* for companies. In our context, regimes can induce new market opportunities for renewable energy sources as well as technologies for energy efficiency. In addition, international regimes can provide common policy instruments in order to establish equal competitive frameworks. One obvious example is emissions trading, which has received general support by industry. Business and industry tend to prefer internationally harmonised, flexible and cost-effective policy instruments (Skjærseth, 2000).

On this basis, we propose that change and differences along the above dimensions have led to change and differences between European and US-based companies. More specifically, we explore the proposition that a regime that progresses beyond the interests of industry, that provides a common understanding of the problem at hand, that induces pressure and grants opportunities will stimulate a proactive strategy among relevant target groups.

Conclusion

In this chapter, an effort has been made to identify critical factors or key conditions that determine the strategies chosen by the oil industry to meet climate-change challenges. Since large multinational oil companies represent important target groups for mitigating climate change, identifying such conditions will provide knowledge of whether and how corporate resistance to climate policy can be overcome. Of particular importance is the extent to which varying climate strategies are the result of company-specific factors, or whether they are located in the political context at national or international levels. Strong empirical support in the former case will point in the direction of corporations operating beyond the reach of climate policy-makers, while the latter case is more hopeful in terms of governance: corporate climate strategies can be affected within existing political frameworks at national and international levels.

The first task in this chapter was to distinguish between different corporate climate strategies. The main challenge is that the public profile of a company, i.e. the rhetoric it uses, may diverge significantly from actual behaviour. Accordingly, there is a real danger of exaggerating differences in corporate strategies. We have tried to deal with this pitfall by identifying indicators that emphasise the companies' activities rather than the rhetoric they use: what have the companies actually done to confront climate change? Regrettably, comparable data on GHG emissions at corporate level are not available. We should also bear in mind that radical changes in behaviour should not be expected at this stage, since climate change is a relatively new issue area. We base our ranking of corporate strategies on a continuum from reactive to proactive.

The next step was to identify the key factors affecting corporate strategy choice. These factors can be identified within a multi-level governance approach. Variations in corporate strategies can be due to variations at corporate, national or international levels. The driving forces at these levels were captured within the framework of three models: the CA, the DP and the IR. The IR model deviates from the former two in that it places more emphasis on capturing the conditions triggering changes in corporate strategies over time.

Before we delve into an empirical scrutiny of these models, we will take a closer look at the specific climate strategies of ExxonMobil, Shell and Statoil.

Notes

1 See, inter alia, Post and Altman, 1992; Roome, 1992; Steger, 1993; Ketola, 1993; Hass, 1996; Ghobadian et al., 1998.
2 In addition, a company may adopt a proactive environmental strategy as a response to societal demand for environmental protection measures expressed by its own shareholders and employees. This dimension can also be extended to cover 'counter demand', e.g. demand for cheap petrol. If climate-change measures mean higher petrol prices, variation in demand for cheap petrol can contribute to an explanation of corporate climate strategies.
3 It is important to emphasise that the purpose here is *not* to explain why climate policy differs and changes within and across countries. Climate policy may be caused by other factors, such as energy-economic circumstances (see Chapter 5).
4 Climate-policy instruments can also be understood in a broader sense, when organisation, information and research and development (R & D) are included.
5 Note that there is no necessary relationship between authoritative force and effectiveness measures in terms of behavioural change.
6 In fact, international regimes produce a chain of consequences, and effectiveness can be measured at different points in this chain: *Output1* refers to joint international commitments; *output2* is

related to domestic policy and implementation of joint commitments; *outcome* points to changes in the behaviour of target groups; and impact refers to the tangible consequences affecting the physical problem at hand.

7   Personal communication with Brian Flannery, ExxonMobil, Irving, Texas, 16 March 2000.

3

The climate strategies of the oil industry

Oil companies want to sell as much oil and gas as possible at the highest possible price. Still, a quick glance at the web pages of Shell, ExxonMobil and Statoil (as well as other US and European-based oil companies) reveals significant differences in their perceptions of climate change. What are the strategies adopted by ExxonMobil, the Shell Group and Statoil on the climate issue? Do they merely use different rhetoric to please their clients, consumers and employees, or is the observed difference of a deeper nature? And to what extent have their climate strategies undergone changes during the last decade?

In this chapter, these companies' strategies are assessed and compared with a main focus on a key set of four indicators: (1) the companies' acknowledgement of the prospective problem of a human-induced climate change; (2) their position with regard to the Kyoto Protocol; (3) self-imposed targets and measures to reduce GHG emissions from their own operations; and (4) the long-term implications of their strategy choice, analysed in terms of the degree of reorientation in their core business areas. With regard to the last indicator, an attempt is made to make qualified judgements regarding the extent to which the strategies of the companies to climate change have implied, and will continue to imply, significant changes in their investment decisions.

The climate strategies of ExxonMobil, the Shell Group and Statoil are assessed in relative terms according to a continuum from 'reactive' to 'proactive' strategies. To the extent that companies acknowledge the climate problem, support the Kyoto Protocol, adopt targets and measures to reduce emissions from

their own operations, and adopt strategies with long-term implications for their mode of operation and business profile, we judge
their choice of climate strategy to be proactive.

The climate strategies of ExxonMobil, the Shell Group and
Statoil are assessed in the three first sections respectively. To place
the companies' climate strategies in a broader context, the more
general environmental policy of the companies is briefly assessed
in an introductory note to each section of the chapter. The
comparison of the strategies chosen by the three companies is
carried out in the last section.

ExxonMobil Corporation

Exxon Corporation started out as Standard Oil in 1882, mainly
as a refinery company. In 1888 it began to internationalise its
downstream assets, and in the 1920s it invested heavily to become
a fully integrated oil company (Estrada et al., 1997). Exxon was
one of the 'seven sisters', the oil cartel that controlled the world
oil trade in the first half of the twentieth century. In the 1970s,
the oil industry was rocked by the Arab oil crisis, and both Exxon
and Mobil escalated exploration and development outside the
Middle East – in Africa, Asia, the Gulf of Mexico and the North
Sea. In November 1999, Exxon and Mobil merged to form
ExxonMobil Corporation. Exxon's takeover of Mobil was the
largest in history: Mobil shareholders own about 30 per cent of
the new company, while Exxon shareholders own about 70 per
cent. ExxonMobil is at present the largest multinational company
in the world, irrespective of sector.[1] The immense size of the
company is indicated by its financial data: in 2000, the company
had a record net income of US$17.7 billion with total revenues
exceeding US$230 billion (ExxonMobil, 2000a). The company
conducts business in gas, oil, coal and chemicals in more than 200
countries.

Until its merger with Mobil in 1999, Exxon was a strongly
hierarchical organisation, with the company's headquarters
playing a major role in its decision-making process. For instance,
all investments exceeding US$1 million needed approval from
headquarters, although operational managers were given a
certain latitude to implement and organise their own programmes
and plans as appropriate to the geographical character of their

areas (Estrada et al., 1997). The merger with Mobil brought about a reorganisation, which has implied a more decentralised structure. With the reorganisation, 'ExxonMobil has entrusted its vast and diverse operations to a slate of business units with global responsibilities. Each company stewards a focused portfolio of operations around the world with a president at the helm and significant authority to run themselves.'[2] Headquarters nevertheless still play a major role in the decision-making process of the company.

ExxonMobil's environmental policy

ExxonMobil places a strong emphasis on excellence in its environmental performance. The 1999 Health, Safety and Environment (HSE) Progress Report, for instance, states that 'without success in these areas, we cannot succeed in operational or financial terms'.

Three elements stand out as fundamental to ExxonMobil's environmental policy.[3] First, the company places a strong emphasis on being in *compliance* with environmental laws and regulations. This does not necessarily mean that ExxonMobil is less likely to implement self-imposed environmental standards exceeding the level of ambition in existing environmental laws. For example, its annual report states that the company should apply 'responsible standards' where environmental laws and regulations do not exist (ExxonMobil, 2000a). As pointed out by Estrada et al. (1997), however, the company does not explain what it considers to be 'responsible standards'. Also, it does not describe or discuss the potential conflict between implementation of such standards and what the company defines as its overarching goal: 'ExxonMobil is committed to being the world's premier petroleum and petrochemical company. Through the execution of long-standing, fundamental strategies that capitalize on our core strengths, the company achieves superior financial and operating results that enhance the long-term returns to our shareholders' (ExxonMobil, 2000b: 1). Nevertheless, laws and compliance are central references in ExxonMobil's statement of its environmental policy.

Second, the *scientific basis* of environmental regulations is a central element for ExxonMobil. It is a stated policy for the corporation to work with government and industry groups to develop environmental laws and regulations that are 'based on sound science'. Thus, ExxonMobil also places a strong emphasis on its own research competence, operating with a stated policy to 'conduct and support research to improve understanding of the impact of its business on the environment, to improve methods of environmental protection, and to enhance its capability to make operations and products compatible with the environment' (ExxonMobil, 2000a).

Third, an important point for ExxonMobil is that environmental policies and regulations should be developed and reviewed with a view to the *broader context* of environmental issues. In particular, policies should be seen in relation to the economic dimensions of environmental protection measures, including a consideration of risks, costs and benefits, and effects on energy and product supply (ExxonMobil, 2000a). Thus, it is also a stated policy for ExxonMobil 'to conduct its business in a manner that is compatible with the *balanced* environmental and economic needs' of the communities in which the company operates (ExxonMobil, 1999a, emphasis added).

To ensure that policy commitments are transformed into 'appropriate' action, Exxon developed its Operations Integrity Management System (OIMS) in the early 1990s (figure 3.1). These activities were stepped up in the wake of the *Exxon Valdez* incident in Alaska in 1989 (see chapter 4). The OIMS, which also is a central part of HSE management after the merger, serves as a structured system for 'identifying and managing [HSE] risks, for compliance with all applicable laws and regulations and for designing and operating facilities to the highest standards' (Exxon, 1999a; see also ExxonMobil, 2001a). It lays out ExxonMobil's HSE requirements for all its business units and also includes a system to maintain accountability and assess how standards are being met. The OIMS requires periodic assessments that are carried out by multidisciplinary teams of experts external to the immediate unit (ExxonMobil, 2000a).

ExxonMobil's climate strategy

ExxonMobil's emphasis on compliance, 'sound' science and consideration of economic impact permeates its approach to the problem of a human-induced climate change. According to ExxonMobil, climate projections are based on 'completely

Figure 3.1 *ExxonMobil's Operations Integrity Management System*

1 Management leader-
ship, commitment and
accountability

2 Risk assessment management
3 Facilities design and construction
4 Information and documentation
5 Personnel and training
6 Operations and maintenance
7 Management and change
8 Third-party services
9 Incident investigation and analysis
10 Community awareness and emergency
preparedness

11 Operations integrity
assessment and
improvement

Source: ExxonMobil (2000c).

unproven climate models or more often on sheer speculation, without a reliable scientific basis'.[4] ExxonMobil's account of the scientific foundation of the climate problem is that it is 'good advocacy' but 'bad science' since 'the facts aren't there' (Flannery, 1999: 4).[5]

ExxonMobil thus opposes the Kyoto Protocol as a 'premature international initiative' that has 'the potential to cause economic harm for most nations, severely impacting some, while doing very little to influence the climate' (Flannery, 1999: 4). Three main arguments underlie ExxonMobil's opposition to the Protocol:[6]

1 *It is too expensive.* ExxonMobil presents an economic analysis that shows that the targeted reduction in fossil-fuel use would mean a 45 per cent increase in petroleum prices and cost an average American family of four about US$2,700 a year. It is maintained that some developed countries probably would have to impose 'significantly higher fossil fuel taxes, rationing, or lifestyle changes such as mandatory car-pooling' (Flannery, 1999: 7).

2 *It is unfair.* The company maintains that projections show that developing countries, including China, Mexico, Brazil and India, will account for almost 70 per cent of total carbon emissions growth from 1990 to 2025. These countries are not included in the Kyoto agreement. This, ExxonMobil maintains, 'raises the question of whether that agreement is fair'.[7]

3 *It will not work.* The company maintains that the warming projected by 2100 would be delayed by only some 10 years as an effect of the Kyoto Protocol and that 'far more onerous emissions reductions would be necessary if climate change proves to be serious'.[8]

Thus, ExxonMobil considers the problem of a human-induced climate change as a 'legitimate concern', but claims that current knowledge of the science and economics of climate change does not warrant significant and mandatory GHG emissions reductions. Rather ExxonMobil argues that if emissions from human activities are altering the global climate, it is a change that is taking place on a long-term time scale. Thus, according to ExxonMobil, we have plenty of time to do more research before policy-makers address the problem. Central to ExxonMobil's argument is that the key to the solution to the climate-change problem is technological development.[9] New technology is seen as the key enabler of combining economic prosperity and environmental protection.[10] ExxonMobil has thus not adopted GHG emissions reduction targets and measures for its own operations.

Exxon, and now ExxonMobil, has given a fair amount of attention to the problem of global climate change in its public relations. The company is publishing booklets and brochures in which its view on the issue is explicated and defended, and the issue is discussed in depth in its annual reports, in HSE progress reports, and in speeches, presentations and press releases that company officials have issued on various occasions. Also, ExxonMobil has been very active in lobbying efforts against governmental GHG regulations in general and US ratification of the Kyoto Protocol in particular. From the establishment of the Washington-based GCC in 1989 – the main vehicle for the fossil-fuel lobby's campaigns against GHG regulation – until its 'deactivation' in 2002, ExxonMobil functioned as a key member. ExxonMobil is also a key member of the American Petroleum Institute (API), whose policy on climate change is an almost verbatim copy of ExxonMobil's. Within the framework of these and other lobbying groups, ExxonMobil has spent a serious amount of money on PR campaigns to influence public opinion and policy-making on the issue. In the 1990s, ExxonMobil reportedly spent more than £700 million financing the GCC[11] (see also chapters 5 and 6).

Being a science and technology-based corporation, ExxonMobil's emphasis on sound science, technology development and economic analyses also constitutes the foundation for the corporation's initiatives on climate change. Five areas of activity – falling within the broad categories of research and no-regrets measures (measures that are justified on other grounds) – are emphasised as core in a responsible path forward:

1 *Conducting scientific research.* ExxonMobil recognises the long-term risk of climate change. To promote a better understanding of the science of climate change, ExxonMobil conducts climate research itself as well as supporting research by others. According to the company itself, ExxonMobil scientists have 'published over 25 papers on climate change in the

peer reviewed literature, often with distinguished academic researchers' (ExxonMobil, 2001b: 4).

2 *Carrying out cost-benefit analyses of proposed responses.* The corporation argues that 'citizens have a right to know the consequences of suggested governmental policies before they are implemented'.[12] Recognising that policy mistakes can be serious, even limiting our opportunity to respond effectively later, ExxonMobil sees it as an important task to undertake such economic analyses.

3 *Encouraging voluntary action.* In 1998, Exxon established a task force to develop a comprehensive global energy management system to improve energy efficiency further at all refineries and chemical plants. In 1999, ExxonMobil had improved energy efficiency by 37 per cent in its refineries and chemical plants over the past 25 years, thus – according to its own estimate – saving an equivalent of a total of 1.7 billion barrels of oil (ExxonMobil, 2001b: 2). ExxonMobil has also made investments in co-generation of heat and power (CHP) facilities, and operates or has interest in over 2,000 megawatts of co-generation capacity worldwide – 'enough to meet the residential needs of a city with 3 million people' (ExxonMobil, 2001b: 2).

4 *Investing in technological research and development.* ExxonMobil invests in technological options exclusively within the framework of its fossil-fuel-based portfolio. Promising technological options (those with the potential to reduce future emissions significantly while meeting energy and economic needs) include fuels and power plants for advanced vehicles such as gasoline–electric hybrids and fuel cells; clean-coal technology for electricity generation; separation and storage of CO_2 emissions; and geo-engineering to remove carbon dioxide directly from the atmosphere (ExxonMobil, 2001b: 8). Currently, investments in 'promising technological options' do not include investments in renewable energy sources. In view of 'their technology limits and excess costs', a business decision was taken many years ago to concentrate on the core energy and petrochemical businesses of the corporation. In fact, ExxonMobil strongly opposes government policies to promote renewables, as clearly expressed by ExxonMobil chairman and CEO Lee R. Raymond in 1996:

'governments should not try to pick winners by subsidizing one alternative fuel over the other or by specifically discriminating against oil-based fuels ... The challenge for us in the petroleum industry is to do what I am doing today – stand up and tell people that oil-based fuels are plentiful, affordable, clean and getting cleaner all the time.'[13] Technical and business developments in renewables are, however, followed closely, and 'if and when relevant technologies allow their commercial utilization in more than niche applications, ExxonMobil is well positioned to consider re-entry' (ExxonMobil, 2001b: 8).

5 *Promoting carbon storage.* Despite its scientifically uncertain long-term effect on the reduction of atmospheric concentrations of CO_2, ExxonMobil pledges to protect and expand forests and to promote soil management where economically justified. Thus, ExxonMobil contributes financially to tree-planting programmes in the US and other countries (ExxonMobil, 2001b; Exxon booklet, undated.)[14]

Particularly when seen in the light of ExxonMobil's relentless efforts to fight down any mandatory regulation of GHG emissions, it is clear that these five activities do not in any way challenge the fossil-fuel portfolio of the company. Barring carbon storage, the activities are all no-regrets measures that are difficult to distinguish from the normal day-to-day activities of a company such as ExxonMobil. Also, their research activity is very much directed towards scientific verification of the claim that the climate does not represent a problem that needs to be addressed by governments. The climate strategy of ExxonMobil, therefore, has no long-term implications for the company's business profile or mode of operation. ExxonMobil adopted its climate strategy when the issue surfaced on the international political agenda in the late 1980s and the company has not changed its strategy since.

ExxonMobil's climate strategy summarised   First, ExxonMobil does not accept that a human-induced climate change constitutes a problem that needs concerted action by governments. While it recognises that climate change is a legitimate (long-term) concern, it also emphasises that the scientific basis is still too uncertain to justify costly action. Thus, ExxonMobil does not accept that its activities – as a fossil-energy company – represent a documented climate risk.

Second, ExxonMobil vigorously opposes the Kyoto Protocol: while ExxonMobil does see its HSE performance as essential for its success in operational and financial terms, the problem of a human-induced climate change is not seen in these terms. On the contrary, the corporation paints a grim picture of the economic consequences of 'premature' policy action (i.e. the Kyoto Protocol) in this area – for the kind of business ExxonMobil is running, as well as for society in general.[15] As we have seen, an important aspect of ExxonMobil's strategy on climate change has been active lobbying against the Kyoto Protocol's ratification in the US Senate. With its deep involvement in both the GCC and the API's activities related to climate change, ExxonMobil has represented a major actor and driving force in the campaign against the Protocol in the US.

Third, ExxonMobil has not adopted explicit GHG emissions reduction targets and measures for its own operations. The measures the company cites as its climate-related initiatives have a strong emphasis on voluntary approaches directed towards the development of new technology (within the framework of its fossil-fuel portfolio) and energy efficiency.

Fourth, as means to deal with climate change, therefore, the efforts undertaken by ExxonMobil have no implications for the mode of operation or the business profile of the company, in either the short or the long term. On the contrary, the strategy is designed to permit business as usual. Finally, the company's climate strategy has remained unaltered since it was first adopted.

In accordance with our discussion of strategy typologies in chapter 2, ExxonMobil's approach to the climate issue bears the distinct characteristics of a reactive strategy.

The Royal Dutch/Shell Group of Companies

The Royal Dutch/Shell Group of Companies is the result of an alliance made in 1907 between the Royal Dutch Petroleum Company and the 'Shell' Transport and Trading Company, plc, whereby the two companies agreed to merge their interests on a 60:40 basis while keeping separate identities. Shell was also one of the 'seven sisters', the influential oil company cartel during the first half of the twentieth century. Before the ExxonMobil merger, Shell was the biggest multinational oil company in the world (Estrada et al., 1997). In 2000 it ranked as the second largest multinational oil company (after ExxonMobil) and the sixth largest multinational company in the world, irrespective of sector.[16] In 2000, the Shell Group had a net income of US$12.7 billion, with total revenues of US$149.5 billion. Currently the company conducts business in more than 135 countries in oil, gas, chemicals and renewable energy sources.

In contrast to ExxonMobil, Shell operated with a highly decentralised company structure from its origin in 1907 until 1995. From 1959 until 1995 the company was organised according to the 'McKinsey-derived matrix structure', which was unusually complex. During this period, 'most decision-making was concentrated at the level of the 100-odd local operating companies' (Neale, 1997: 96). In 1995, however, Shell began a process of reorganisation, which has led to a stronger degree of centralisation. Today, Shell is organised in five global functional core businesses that, while independent, comply with the same set of business principles. Shell's corporate headquarters – Shell International – are located in London, but the Dutch branch of the company (the Royal Dutch) owns 60 per cent of the assets.

Shell's environmental policy

Compared to ExxonMobil, the Shell Group has a more principled approach to environmental protection issues. The company has had a written environmental policy since 1969. Its environmental commitment is included and specified in Shell's General Business Principles, which have existed in written form since 1976. Within this principled framework, the company published its first *Guidelines on Health, Safety and the Environment* in 1977 (Estrada et al., 1997). The Shell Group's HSE commitment emphasises the following principles:[17]

- protecting both the environment and human populations;
- using material and energy efficiently;
- developing energy resources that are consistent with these aims;
- public reporting on performance;
- promoting best practice in its industries;

- prioritising HSE as any other critical business activity;
- promoting a culture in which Shell employees share this commitment.

In its environmental policy, Shell seeks to transform these general principles into a practical policy. Thus, it is mandatory for every Shell company to:

- have a systematic approach to ensure compliance with the law and continuous performance improvement;
- develop clear targets for both performance and reporting;
- include HSE performance in the appraisal of all staff;
- reward the staff accordingly;
- subject all Shell's contractors and cooperative partners in joint ventures to the same environmental policy and standards.

Shell's global performance data are published in its HSE Progress Reports and have been externally verified since 1997. The verification is focused on 12 HSE parameters including CO_2 emissions, methane emissions, global warming potential (in million tonnes CO_2 equivalents), flaring, sulphur dioxide emissions (SO_2) and nitrogen oxide emissions (NO_x) (Shell, 2000a). Having started the development of a systematic approach to HSE management in 1998, all Shell Operating Units now have their environmental management systems certified against recognised, independent system standards, such as ISO 14001 or the European Union's Eco-Management and Audit Scheme (EMAS).

Shell's climate strategy
In its description of the climate problem as 'the most controversial and pressing environmental issue we face', Shell fully acknowledges the problem of a human-induced climate change[18] (Shell, 1998b). Furthermore, immediately preceding the start of climate negotiations in Kyoto in 1997, Shell announced its support for 'prudent precautionary action to reduce man's impact on the global climate'.[19] The company has later explicitly expressed its support of the Kyoto Protocol.[20]

Shell's position with regard to governmental action to reduce GHG emissions, however, has changed rather abruptly from a reactive stance until 1996 to the current proactive strategy. As late as July 1996, Shell, together with 119 chief executives and chairmen of mainly oil, coal and car companies, added its signature to a letter to the US president arguing that 'the US should not agree to any of the three proposed protocols presently on the negotiating table. Your leadership on this issue is critical to assuring a continuing strong US economy' (Leggett, 1999: 246). BP was the only one of the major multinational oil companies that did not sign the letter. As we have seen, statements to the opposite effect, proclaiming support of precautionary action, started to emerge from Shell in 1997. In April 1998, the company's exit from the fossil-fuel lobby became official with its decision to withdraw from the GCC because of its disagreement with the coalition's strategy on the Kyoto Protocol: 'Following Kyoto it became clear that the respective views of the Shell companies and the GCC were too far apart ... The GCC is actively campaigning against legally binding targets and timetables as well as ratification by the US government. The Shell view is that prudent precautionary measures are called for' (Shell, 1998b). Thus, Shell's position started to change in 1997, and the company's turn-around on the climate issue was completed by 1998, after the adoption of the Kyoto Protocol.[21]

In September 1998, the Shell Group announced its aim to reduce GHG emissions from its own operations by 10 per cent from their 1990 levels by 2002. Shell also aims at continuing to exceed the Kyoto target by 2010.[22] While Shell's GHG emissions went up in 2001, the company reported that it was still on track to reach the emissions reduction target by the end of 2002 (Shell, 2001).

Shell's commitment on climate change, outlined in the 1998 and 2000 Shell reports (1998b; 2000a), envisages six strategies to achieve the Group's long-term climate change commitment:[23]

1 *Seek market solutions.* In January 2000, Shell launched an internal GHG emissions trading system called the Shell Tradable Emission Permit System (STEPS). Businesses representing 30 per cent of the GHG emissions from the Shell Group's operations are now using tradable emissions permits to help meet their self-imposed emissions targets.[24] In this system, participants are rewarded for reducing their emissions wherever the cost is lower than the price of a GHG emission permit. Shell also has as part of its strategy the provision of practical support in the development of national and interna-

tional emissions trading systems. In February 2002, Chris Fay, former head of Shell UK, was appointed by the British government to promote the UK emissions trading scheme in which Shell participates. Finally, the company also seeks opportunities to invest in projects using the implementation mechanisms embedded in the Kyoto Protocol. Shell invests some US$3–4 billion per year in non-Annex B countries in all of its five core business areas.[25] The Shell Clean Development Mechanism (CDM) Demonstration Programme analysed and assessed the impact of CDM for eight projects. In 2001, Shell created an Environmental Products Trading Team within the Shell Trading organisation. This group is also responsible for exploring CDM opportunities for the Shell Group.[26]

2 *Make appropriate business decisions.* As nations choose different ways to meet their Kyoto targets, such actions will put a cost on carbon emissions that will influence the investment decisions of Shell companies. To meet this challenge, Shell includes the effect of a possible carbon cost in their investment decisions for new projects that could produce emissions over 100,000 tonnes a year of CO_2. It will also investigate ways to reduce carbon emissions, for example by improving energy efficiency, using low-carbon fuels, and carbon sequestration or carbon removal from the atmosphere.

3 *Reduce own emissions.* The company will continue to invest in energy efficiency in its operations and stop the continuous disposal of unwanted gas during oil extraction (by venting and flaring) as early as possible. The Group's target is to halt continuous venting by 2003 and continuous flaring by 2008.

4 *Help customers reduce their GHG emissions.* This aim will be met by increasing the availability of fuels with a lower carbon content; by offering renewable energy choices; and through work on new innovative technologies such as hydrogen.[27]

5 *Improve understanding.* Shell seeks to expand support for research into climate change and its impacts, and to promote a deeper understanding of the 'wells-to-wheels' concept, which enables a comparison of the efficiency of different fuel and engine combinations. It also seeks to take active part in public policy debates at the national and international level directly, and through industry bodies and international organisations. More generally, Shell aims at contributing to a deeper under-

standing of the issues through the development of its social investment programmes.

6 *Develop reporting and verification systems.* Measurement is seen as 'fundamental' to many of the initiatives the company has proposed. Together with BP, Shell has been instrumental in developing the Greenhouse Gas Protocol Initiative, initiated by the World Business Council for Sustainable Development (WBCSD) and the World Resources Institute (WRI) in 1998, aimed at promoting internationally accepted GHG accounting and reporting standards for companies. BP and Shell have also been front runners in GHG emissions verification by third parties (Loreti et al., 2001). As a consequence of BP's and Shell's activities in this area, a growing number of companies have shown interest in emissions verification. The companies' verification systems have also been important for public authorities' requirements to environmental reporting.

Shell's initiatives on climate change have implied significant changes in its mode of operation and business orientation. First, a central aspect of Shell's strategy on climate change is 'decarbonisation' of fossil fuels – both through improved technology and through a switch from coal to oil to gas (see, for instance, Fay, 1997). In line with such an endeavour, Shell announced in 1997 that the group's coal assets were under strategic review. The sales of its coal assets were completed in 2000. With the 1997 announcement, Cor Herkströter, chairman of Royal Dutch Shell, also reportedly 'made plain he backs world moves towards "decarbonisation" – getting rid of one of the biggest causes of the greenhouse effect'.[28]

The most visible result of Shell's position on climate change is the 1997 restructuring, when Shell International Renewables was established as a fifth core business activity. The purpose of this restructuring was to consolidate the Group's activities in solar power, biomass (wood-based) power and forestry, and it underscores the Group's strategic direction, which is 'to provide energy and develop resources efficiently, responsibly and profitably in order to help meet the world's growing needs, and to do so in a way that contributes to sustainable development'.[29] Shell maintains that renewables will constitute the main energy source of the future.[30] With the establishment of Shell International

Renewables, Shell aimed to capture a 10 per cent share of the renewables market before 2005. Shell planned to invest more than US$0.5 billion over a five-year period from 1997 in renewable resources. This, however, represents only a small fraction of the company's investment, for instance, in oil exploration.

As noted in chapter 2, it is difficult to assess the cause–effect relationship between a proactive strategy to climate change and a strategy of decarbonisation of fuels and introduction of renewables. Shell itself has presented its divestment in coal as an integral part of its climate strategy. While these events all took place in 1997, and hence at approximately the same time, Shell's shift in climate strategy was nevertheless announced before its decision to divest its coal assets and invest in renewables (although the company remained a member of the GCC until April 1998). On the basis of the chronological order of these events and Shell's own stated motivation, Shell's actions on coal and renewables are thus treated as effects of a new strategy in this analysis, rather than as causes of its climate strategy.

Moreover, with regard to investments in renewable energy sources, Shell emphasises business opportunities in this area as its main motivation. Shell's investment in renewables should thus also be seen as a response to the one main challenge, as Shell sees it, of finding commercial paths to new energy futures.[31] The commercial aspect, therefore, is one central focus, and reportedly a main motivation for Shell's position on and investments in environmental protection in general and the climate-change issue in particular. Shell invests in greenhouse gas reductions and renewable resources because they see such an investment as a business opportunity.[32]

Shell's emphasis on the commercial aspects of environmental protection and sustainable development is also reflected in a strong support for and emphasis on the development of 'achievable targets', primarily through market-based instruments and measures to reduce emissions.[33] In 1997, then chairman and chief executive of Shell UK Ltd, Chris Fay, outlined four groups of instruments – in prioritised order – that could be used to reduce and change energy-consumption patterns:

• ending subsidies for coal production in countries like Germany, China and India;

- tackling other artificial pricing structures that encourage high-carbon fuels;
- ensuring that there are competitive electricity markets so that new technologies can find market niches;
- looking at other consumption-dampening mechanisms such as carbon taxes, tradable carbon permits and so on.

According to Fay, it is important to ensure that measures to reduce consumption 'are not disguising revenue raisers', that such measures 'encourage competitive markets, not distort them', and that they thus 'must be aimed at pushing the energy industry along the development path that is already clearly evident – toward lower carbon content'.[34]

Shell's climate strategy summarised First, Shell acknowledges the problem of a human-induced climate change and bases its climate strategy on the precautionary principle. Shell thus acknowledges that its activities in the coal, oil and gas industry represent an environmental risk.[35]

Second, Shell explicitly supports the Kyoto Protocol and has been a front runner together with BP in promoting reporting, verification and emissions trading.

Third, Shell has adopted targets and measures to reduce GHG emissions from its own operations. The company aimed at a 10 per cent reduction from 1990 levels by 2002. One important instrument to achieve this goal was the establishment of Shell's internal emissions trading scheme. Another important measure adopted by Shell was to calculate an anticipated carbon cost in investment decisions for new projects that would emit more than 100,000 tonnes of CO_2 a year.

Fourth, Shell's climate strategy has had and will continue to have implications for the business orientation of the company: Shell divested its coal activities in 2000 and was reorganised in 1997, with the establishment of Shell International Renewables as a fifth core business area. The primary motivation for this establishment was the market potential judged by Shell to be associated with renewables and a future renewables market. Perhaps equally important for Shell's future business profile is the Group's decision to include anticipated carbon costs in its investment decisions for new projects. Thus, Shell has reviewed and revised its business profile in response to the climate problem.[36] It is interesting to note that Shell's strategy on this issue is similar to that chosen by the other European oil giant, BP (Rowlands, 2000).

In accordance with our discussion of strategy typologies in chapter 2, Shell's climate strategy can thus be characterised as proactive both in relation to the criteria selected here and in comparison with ExxonMobil's. Shell's approach to climate strategy underwent a complete reversal during the 1990s, however, from a reactive approach until 1996 to the current proactive approach that the company has adopted since 1997/1998.

Statoil Corporation

Statoil is a youngster compared to ExxonMobil and Shell. In 1972 the Norwegian authorities established Statoil as a fully state-owned oil company for the exploration, production, refining and marketing of the petroleum resources found on the Norwegian continental shelf. Rapid growth ensued in the 1980s, as Statoil was given the responsibility of running operations at Gullfaks in 1981 and Statfjord in 1987. During the 1990s, the company gradually expanded its international upstream operations and currently operates in 25 countries. With its operating revenues in 2000 of 208 billion Norwegian kr. (US$23.5 billion), Statoil is a dwarf among giants on the international scene. Its net income in 2000 was 11.3 billion Norwegian kr. (US$1.3 billion).[37] The company, however, is the world's second largest seller of crude oil and a significant supplier of natural gas to Europe. The company ranks as the biggest retailer of petrol and other oil products in Scandinavia. Moreover, Statoil was responsible for managing the Norwegian state's direct financial interest (SDFI) on the Norwegian continental shelf from its establishment in 1985 until 2001. The Statoil and SDFI combined portfolio ranks as the fourth largest of OECD oil companies with regard to both oil and gas production and oil and gas reserves (after ExxonMobil, BPAmoco/Arco and the Shell Group) (Statoil, 1999b).

Statoil was a fully state-owned company from its establishment in 1972 until April 2001, when Norwegian authorities decided to privatise Statoil partially and list the company on the stock exchange. Initially, only 18.2 per cent of the company was sold to private shareholders, but the Norwegian parliament will over time reduce the state's ownership to two-thirds of the shares. Norwegian authorities also decided to sell approximately 20 per cent of the value of the SDFI, of which Statoil bought 15 per cent. A new state-owned company, Petoro, was established to manage the remaining shares of the SDFI. Statoil, however, remains responsible for selling the SDFI's oil and gas volumes.[38]

Statoil's environmental policy

Statoil's systematic approach to environmental issues is of relatively recent origin: the early 1990s. In 1991, the company adopted the 16-point charter for sustainable development drawn up by the ICC. Also in 1991, a project to identify environmental challenges and to strengthen the company's work on environmental issues was initiated. An environmental department of five people was established, obliged to report to the corporate management (Estrada et al., 1997). Today, the management system for HSE forms an integrated part of Statoil's total management system.

The company also issued its first environmental report in 1991 – *Responding Actively to Environmental Challenges* – emphasising humankind's responsibility for the environment and the necessity of keeping development within the bearable limits of nature in order to preserve the common environment. Today, under the slogan 'A high performance in HSE has a value in itself', Statoil adopts an 'ethical commitment to preserve human life and health, protect the environment and safeguard material and financial interests as well as our reputation in all circumstances'.[39]

A key element in Statoil's HSE management system is registration, reporting and assessment of relevant data. Statoil has nine group-wide HSE performance indicators, five of which are related to the external environment: oil spills, CO_2 emissions, NO_x emissions, energy consumption and the waste recovery factor. These indicators are reported annually for all Statoil-operated activities (oil spills are reported quarterly) (Statoil, 2000).

Statoil's investments in environmental measures, however, are pragmatically spent on research, installation of new machinery and upgrading of old equipment (Statoil, 1998). The research effort is primarily focused on an increased knowledge about emissions from its own operations and associated health and environ-

mental effects, and the development of products, processes and
technology to reduce emissions and adverse effects.

Statoil's climate strategy
The issue of global climate change has been given attention ever
since the publication of Statoil's first environmental report in
1991. Today Statoil's slogan is 'The issue for Statoil is not
whether the world faces a climate problem, but how it can be
overcome.'[40] Thus, Statoil seems to accept readily the scientific
basis of the climate-change problem. Statoil explicitly states that
it will make active efforts to stay up to date on developments in
scientific knowledge about the greenhouse effect, and that the
company will continue, together with others, to contribute to an
understanding of the social, economic and competitive impacts of
climate policies aimed at the petroleum industry and the energy
market. Moreover, it will actively participate in an open collaboration with the authorities to find effective solutions for abating
GHG emissions.[41]

Statoil has announced its support for the Kyoto Protocol, and
says that it provides a good initial basis for global cooperation on
a rational climate policy (Statoil, 1997). The company finds the
reduction objective agreed upon in Kyoto ambitious, but asserts
that it is achievable if industry and the authorities cooperate in
identifying realistic measures (Statoil, 1997). Accordingly, Statoil
strongly supports the Kyoto mechanisms, which 'will make it
easier to come up with cost-effective solutions and help to ensure
that more and better action is taken globally' (Statoil, 1997).
Since 1991, before the Earth Summit in Rio and the establishment
of the UNFCCC, Statoil has emphasised the importance of flexibility mechanisms in environmental management (Statoil, 1992:
4). Such programmes are especially crucial to Statoil and the rest
of the Norwegian petroleum industry, given the existing levels of
efficiency and high cost of further emissions reductions on the
Norwegian shelf relative to other petroleum-extracting provinces
(Statoil, 1998: 4).

Statoil's corporate level goal is 'zero' emissions that cause
lasting damage or have a negative impact on the environment
(Statoil, 1999a: 33). In practice, carbon dioxide, nitrogen oxide
and volatile organic compounds (VOC) emissions cannot be
entirely eliminated. The exact implications of this 'goal', there-

fore, are difficult to judge. In 1997, Statoil adopted a target to cut CO_2 emissions by 30 per cent over the next decade (by 2010) relative to a 'business as usual scenario' (the level emissions would reach with the currently existing technology and practice) (Statoil, 1998: 2). In 2000, this target was modified. The current target is 'to trim 1.5 million tonnes of carbon dioxide equivalent from its annual greenhouse gas emissions by 2010 compared with "business as usual" based on 1997 technology' (Statoil, 2000). This does not mean, therefore, that Statoil intends to cut its CO_2 equivalent emissions by 1.5 million tonnes annually in absolute terms, but rather to cut 1.5 million tonnes of CO_2 equivalent emissions annually as compared to what the levels would have been with 1997 technology. It is indicated that this goal would correspond to a cut of 'roughly 15% from present *forecasts* for Statoil's share of global emissions in 2010' (Statoil, 2000, emphasis added). Again, therefore, the target is stated in relative terms – relative to an unspecified 'business as usual' baseline scenario. This is a highly ambiguous goal.

International operations are expected to account for a large proportion of the planned emission cuts. Also, energy saving and stopping permanently lit flares will provide major emissions reductions, according to the Statoil environment vice-president, Knut Barland.[42] The focus on technological innovation as a means to fulfil its reduction target became clear when the company revised its goal. On that occasion, Knut Barland commented that the level of ambition in the target was reduced because 'new knowledge has shown that the effect of innovative technology is lower than expected'.[43]

While Statoil does report its CO_2 emissions, it does not report its CO_2 *equivalent* emissions – the measure in which its reduction target is stated (which is a common measure for all GHG emissions, thus also including emissions of other GHGs such as methane and nitrous oxide [N_2O]). Therefore it is difficult to judge whether the company is on track to achieve its goal. Its CO_2 emissions, however, increased by 1.1 million tonnes from 2000 to 2001 (Statoil, 2001). In this perspective, Statoil's general corporate goal of 'zero' emissions that cause lasting damage seems rather meaningless, with little or no practical implications for the operations of the company.

In 1997, Statoil launched its CO_2 programme, a three-year project to identify ways of reducing emissions of this GHG. The main emphasis in the 600 million Norwegian kr. programme was on the development of new technology, making better use of existing solutions, and carbon storage. Thus, Statoil's initiatives on climate change are based well within the framework of their fossil-fuel portfolio.

Statoil's investment in renewables has been modest. In 1997, the company invested in Biovarme, a bioheat company, and started 'Three steps ahead for environmentally sound energy', which maps out investment possibilities in renewables (Greenpeace International, 1998: 54). Also, Statoil is exploring opportunities for producing and marketing renewable forms of energy, such as wind power and biomass (Statoil, 1998: 12). The company is currently collaborating with Norske Skog in the production and sale of biofuels, which are expected to reach an annual production of 8,000 tonnes (Statoil, 1999a: 36). Statoil, along with other oil companies, is also collaborating on fuel-cell research with auto and fuel-cell manufacturers (Statoil, 1999a: 35). Thus, unlike, for instance, BP and Shell, Statoil has not devoted any resources to solar energy research, and in April 1998, Statoil's HSE director reportedly stated that 'renewable energy is not our business' (Greenpeace International, 1998: 54). This is illustrated by the company's decision in January 2000 to withdraw from a joint venture project on wind power in Norway because it could not see any future profitability in the product.[44]

Today, Statoil exports produced gas to continental Europe. The company has in recent years, however, explored the possibility of participating in natural-gas-based electricity generation schemes at home and abroad. For Statoil the prospect of integrating further down the gas chain is considered important for optimising the value of natural gas. Accordingly, Statoil, Norsk Hydro and Statkraft – the largest Norwegian supplier of hydropower – established Naturkraft in 1994. The idea is to produce electricity derived from natural gas and sell it in combination with hydropower. For this purpose, Naturkraft has been working for political and public acceptance of two gas-fired power plants on the west coast of Norway (see chapter 5). The plan has met with considerable resistance from the Norwegian environmental movement, which finds the resulting increase in Norwegian CO_2 emissions unacceptable, not least after the advent of the Kyoto Protocol. However, after a long and polarised political debate (in which one government resigned on this issue), the initiators have achieved a concession to build the plants. The plants have not yet been built, however, mainly because costs are too high and electricity prices too low for the business to be profitable.

Statoil's climate strategy has been developed gradually since 1991. Thus, the strategy has undergone incremental changes during this period. Given the high degree of ambiguity in the company's approach to climate change, however, we cannot say that Statoil's approach has moved towards a larger degree of 'proactivism'. On the contrary, given the company's revision of its CO_2 emissions reductions target towards a lower level of ambition, Statoil's climate strategy has moved back and forth since 1991, and has, above all, remained equally ambiguous throughout the period.

Statoil's climate strategy summarised   First, Statoil accepts that human-induced climate change represents a problem that requires concerted action by governments. Second, the company declares its support for the Kyoto Protocol.

Third, Statoil has adopted a self-imposed GHG emissions reduction commitment. This however, is highly ambiguous, since the reduction goal is stated not in absolute terms but rather in relation to an unspecified 'business as usual' baseline. Statoil plans to implement this target by focusing on technological options to reduce emissions, such as enhanced energy efficiency, technological innovation and carbon storage. Its climate initiatives, therefore, do not challenge the company's fossil-fuel portfolio. This is particularly evident with regard to the circumstances around the modification of Statoil's CO_2 reduction target in 2000.

Fourth, even though Statoil clearly acknowledges the climate problem, supports the Kyoto Protocol and has adopted self-imposed emissions reductions, the level of commitment in this strategy may nevertheless be characterised as moderate. The strategy that Statoil has chosen on the climate issue has no significant (long-term) implications for the company's business orientation.

Finally, while the strategy has developed incrementally since

1991, we cannot trace any significant changes in Statoil's approach towards either a more proactive or a more reactive strategy. Rather, the company's climate strategy has moved back and forth and remained equally ambiguous throughout the period.

In terms of our discussion of strategy typologies in chapter 2, therefore, Statoil seems to lie between a reactive and a proactive strategy, as an 'intermediate' between ExxonMobil and Shell. It has adopted a proactive rhetoric on the issue, but the strategy and choice of means to transform the rhetoric into action are less substantial than Shell's.

Comparison of the climate strategies of the three companies

There are striking differences in the climate strategies adopted by the three companies, particularly between ExxonMobil and Shell. And these differences go far beyond mere rhetoric. In many respects ExxonMobil and Shell represent opposite extremes, with ExxonMobil on the reactive end of the continuum and Shell on the proactive end. Statoil has adopted a strategy that lies in the middle of this spectrum, with similarities to both ExxonMobil and Shell.

Above, we have assessed the climate strategies of the three companies in terms of four main indicators: their acknowledgement of the problem of a human-induced climate change; their position with regard to the Kyoto Protocol; the adoption of targets and measures to reduce GHG emissions from their own operations; and the level of ambition and commitment in their strategies, judged in terms of the degree of reorientation in core business areas implied by the strategies. In addition, we have assessed the degree to which the companies' climate strategies have undergone changes during the last decade.

Acknowledgement of problem and position on the Kyoto Protocol

Of the three companies, ExxonMobil is the most reluctant in its acknowledgement of a prospective human-induced climate change. The company acknowledges that the possibility of human-induced climate change is a 'legitimate concern', but claims it is far from a scientifically established fact. It does not accept that the problem is sufficiently scientifically substantiated to legitimise costly policy regulation. Accordingly, ExxonMobil is also explicitly opposed to the Kyoto Protocol. The Shell Group and Statoil, on the other hand, both acknowledge the climate problem as a real problem requiring concerted action by governments. Both corporations explicitly support the Kyoto Protocol.

Self-imposed GHG emissions reduction targets and measures
Given ExxonMobil's reluctant acknowledgement of the climate problem as a 'legitimate concern' and its explicit opposition to the Kyoto Protocol, it is no surprise that it has not adopted any voluntary targets for GHG emissions control or reduction for its own operations. On the contrary, its position is that if there indeed is a climate problem, it is a long-term problem to which there is plenty of time to develop appropriate responses.

Both the Shell Group and Statoil have announced GHG emissions reduction targets for their own operations. Shell adopted an aim of reducing GHG emissions by 10 per cent from their 1990 levels by 2002. According to its 2001 reporting, the company was set to succeed in achieving this aim, even though its GHG emissions rose during 2001. Statoil, on the other hand, adopted in 1997 a highly ambiguous target of a 30 per cent reduction in CO_2 emissions over the next decade (by 2010) relative to the level anticipated with currently existing (1997) technologies and practice. Moreover, in 2000, this target was modified to an equally ambiguous target of an annual reduction of 1.5 million tonnes of CO_2 equivalent emissions by 2010, also as compared to anticipated 'business as usual' emissions levels (with 1997 technology). While the target was stated in terms of CO_2 equivalents (thus including other gases than CO_2, for instance, methane and N_2O), Statoil does not report its CO_2 equivalent emissions. The CO_2 emissions of the company, however, rose by 1.1 million tonnes from 2001 to 2002. It is therefore difficult to judge whether the company's voluntary reduction target has any practical implications for Statoil's operations.

The most important measures adopted by Shell to implement its reduction target are the establishment of an internal scheme for emissions trading and a revision of its investment decision procedure to include a calculated carbon cost for future projects. Statoil's main measure to implement its target is technological innovation.

Reorientation in business areas

ExxonMobil's current position does not involve strategies explicitly designed to meet a climate problem. Its response to climate change is thus currently heavily dominated by research activities – conducted by itself, in cooperation with other research institutions or through the provision of funds to build 'external capacity'. Other activity in response to this problem mainly consists of technology development to improve energy efficiency and to decarbonise fuels. ExxonMobil does not publish its GHG emissions. Currently, therefore, ExxonMobil's strategy is more or less 'business as usual' and does not show any sign of commitment that might have long-term implications for its business orientation. This strategy is combined with vigorous efforts to defeat any mandatory control on GHG emissions (see chapters 5 and 6).

The Shell Group has also approached the problem by improving energy efficiency and decarbonisation of fuels – for instance, through increases in the provision of natural gas supplied through liquefaction, pipelines and conversion to liquid fuels. In contrast to ExxonMobil, however, Shell's decarbonisation strategy has had implications for its business orientation: in 2000, Shell divested its coal assets. With its investment in renewable energy resources (photovoltaics and biomass energy), the Shell Group, moreover, has adopted a strategy that has already implied a reorganisation of the Group and a reorientation in its business profile. Finally, Shell has changed its routines for investment decisions for new projects to include anticipated costs associated with significant CO_2 emissions. In contrast to ExxonMobil, therefore, it may be argued that Shell's approach to climate change will have much more significant long-term implications in terms of a business reorientation and thus reflects a higher level of ambition.

While Statoil's rhetoric on the climate issue is similar to Shell's (the acknowledgement of the problem, support for the Kyoto Protocol and adoption of voluntary reduction commitments), the company's actions on this issue are more similar to ExxonMobil's. Statoil's strategy towards climate change is largely one of research and technological innovation. To the extent that the company intends to honour its self-imposed reduction target, the reduction will primarily come as a result of technological innovation. This is particularly demonstrated by the circumstances under which Statoil modified its GHG emissions reduction target in 2000. The company has also made no major investments in renewable energy sources. Its climate strategy, therefore, seems to a much lesser extent than Shell's to imply significant investment changes or corporate reorientation. The strategies of the three companies are summarised in table 3.1 below.

Changes in strategies since the start of the 1990s
Of the three companies, only Shell's strategic approach to climate change has undergone a marked and significant change. Shell changed its strategy rather abruptly from a reactive approach until 1996 to a proactive one from 1997/1998. Statoil's climate strategy has been developed incrementally since 1991, but has moved back and forth and has been equally ambiguous throughout the period. ExxonMobil's climate strategy has remained unaltered since it was first adopted in the late 1980s.

Table 3.1 *Summary of the climate strategies of ExxonMobil, the Shell Group and Statoil*

Company	Acknow-ledgement of problem	Position on the Kyoto Protocol	GHG emission reduction target and measures[a]	Reorientation in business areas
ExxonMobil	Reluctant: 'legitimate concern'	Explicit opposition	No	Low: no implications for mode of operation or business orientation
Shell Group	Yes	Explicit support	Yes	High: potentially significant long-term implications for mode of operation and business orientation
Statoil	Yes	Explicit support	Yes, but ambiguous	Medium: shift of emphasis and change in routines, but no long-term implications for mode of operation and business orientation

Note: [a] Beyond 'no regrets'.

Our assessment and comparison of the three companies' climate strategies thus shows that it is not only the rhetoric that divides the companies in this issue area. Shell's climate strategy is supported by a set of actions that already have implied a significant reorientation in the company's business profile, and will continue to have implications for the future. In the next chapter, we explore the extent to which differences between the companies themselves can account for these observed differences in their climate strategies.

Notes

1 Fortune, 2002, Global 500. Source: www.fortune.com (accessed 31 May 2002).

2 Shelley Moore, 'What's new and better about ExxonMobil?', *Lamp* (ExxonMobil publication), Autumn 1999. Source: www. exxon.mobil.com/news/publications/fall99_lamp/04newandbetter. html.

3 Personal communication with Brian P. Flannery and Gary F. Ehlig, ExxonMobil Corporation, Irving, Texas, March 2000.

4 Booklet issued by Exxon Corporation: 'Global Climate Change: Everyone's Debate', undated. Source: www.exxon.com/exxoncorp/ overview/viewpoint/global_climate/global/globe1.htm (accessed 23 August 1999).

5 See also 'Remarks by Dr. Brian Flannery on climate change', 11 March 1999. Source: www.exxon.com/exxoncorp/news/speeches/ speech_031199.html (accessed 12 October 1999).

6 Personal communication with Brian P. Flannery and Gary F. Ehlig, ExxonMobil Corporation, Irving, Texas, March 2000.

7 Booklet issued by Exxon Corporation: 'Global Climate Change: Everyone's Debate', undated. Source: www.exxon.com/exxoncorp/ overview/viewpoint/global_climate/global/globe1.htm (accessed 23 August 1999).

8 Booklet issued by Exxon Corporation: 'Global Climate Change: Everyone's Debate', undated. Source: www.exxon.com/exxoncorp/ overview/viewpoint/global_climate/global/globe1.htm (accessed 23 August 1999).

9 Personal communication with Brian P. Flannery and Gary F. Ehlig, ExxonMobil Corporation, Irving, Texas, March 2000.

10 Booklet issued by Exxon Corporation: 'Global Climate Change: Everyone's debate', undated. Source: www.exxon.com/exxoncorp/ overview/viewpoint/global_climate/global/globe1.htm (accessed 23

August 1999); Flannery, 1999; ExxonMobil press release dated 18 September 1999, 'Mobil report says time, technology and global participation needed to stabilize atmospheric CO_2 concentration'; ExxonMobil press release dated 15 February 2000, 'ExxonMobil Executives: oil industry technology can deliver economic prosperity and a cean environment'; ExxonMobil, 'Global climate change: ExxonMobil views', April 2001. Source: www.exxon.mobil.com (accessed 27 May 2002).

11 *Guardian*, 5 April 2002: 'Oil giant bids to replace climate expert'.

12 Booklet issued by Exxon Corporation: 'Global Climate Change: Everyone's Debate', undated. Source: www.exxon.com/exxoncorp/ overview/viewpoint/global_climate/global/globe1.htm (accessed 23 August 1999).

13 Lee R. Raymond, 'Energy, the economy, and the environment: moving forward together', speech at the Economic Club of Detroit, 6 May 1996. Source: www.exxon.com/exxoncorp/news/speeches/ speech_050696.html (accessed 12 October 1999).

14 Booklet issued by Exxon Corporation: 'Global Climate Change: Everyone's Debate', undated. Source: www.exxon.com/exxoncorp/ overview/viewpoint/global_climate/global/globe1.htm (accessed 23 August 1999).

15 Personal communication with Brian P. Flannery and Gary F. Ehlig, ExxonMobil Corporation, Irving, Texas, March 2000.

16 Fortune, 2002, Global 500. Source: www.fortune.com (accessed 31 May 2002).

17 Source: www.shell.com/royal-en/content/0,528,25583-51073,00.html (accessed 19 March 2001).

18 Personal communication with Gerry Matthews, Shell International, Washington, DC, March 2000.

19 Phil Watts, 'Contributing to sustainable development', statement issued 18 November 1997. Source: www.shell.com/library/ speech/0,1525,2302,00.html (accessed 22 November 1999).

20 See, for instance, 'Progressing towards sustainable development – Shell exploration and production's commitment in action', Shell publication issued 2 November 2000. Source: www.shell.com (accessed 31 May 2002).

21 Personal communication with Gerry Matthews, Shell International, Washington, DC, March 2000.

22 See, for instance, 'Taking action on sustainable development: climate change'. Source: www.shell.com/values/content/0,1240, 1215-3397,00.html (accessed 10 January 2000).

23 Source: www.shell.com/values/content/0,1240,1215-3397,00.html (accessed 10 January 2000); personal communication with Gerry Matthews, Shell International, Washington, DC, March 2000.

24 'Shell Tradable Emission Permit System: an overview', issued 27 January 2000, available at www.shell.com/download/steps/steps.pdf.

25 In the Kyoto Protocol, the quantified emission limitation or reduction commitments by the parties are listed in Annex B of the agreement. Thus, parties that have taken on such commitments are often referred to as Annex B countries. In the UNFCCC, the parties that have taken on the commitments defined in Article 4.2 of the Convention are listed in Annex I of the Convention and are thus often referred to as Annex I countries.

26 Source: www.shell.com.

27 Mark Moody-Stuart, 'The importance of the Kyoto mechanisms for sustainable development and business', speech issued 8 October 1999. Source: www.shell.com/library/speech/0,1525,4407,00.html (accessed 22 November 1999).

28 *Guardian*, 19 November 1997, 'Shell beefs up green moves'. Source: www.globalpolicy.org/finance/alternat/carbon/ct11_19.htm (accessed 31 May 2002).

29 Jeroen van der Veer (group managing director, Royal Dutch/Shell Group) and Jim Dawson (president, Shell International Renewables), 'Shell International Renewables: bringing together the Group's activities in solar power, biomass and forestry', press conference, London, 6 October 1997. Booklet published by the Shell Group.

30 *Guardian*, 19 November 1997, 'Shell beefs up green moves'. Source: www.globalpolicy.org/finance/alternat/carbon/ct11_19.htm (accessed 31 May 2002).

31 Mark Moody-Stuart, 'The importance of the Kyoto mechanisms for sustainable development and business', speech issued 8 October 1999. Source: www.shell.com/library/speech/0,1525,4407,00.html (accessed 22 November 1999).

32 Jeroen van der Veer, (group managing director, Royal Dutch/Shell Group), 'Sustainable solutions support sustainable business', speech issued 26 May 1999. Source: www.shell.com/library/speech/0,1525, 3893,00.html (accessed 22 November 1999).

33 Personal communication with Gerry Matthews, Shell International, Washington, DC, March 2000; Chris Fay (chairman and chief executive, Shell UK Ltd), 'Achievable targets needed', speech at the CBI Panel Debate, 'Climate Change – a Taxing Business?', at the CBI National Conference, Birmingham, 10 November 1997. Source: www.shell.co.uk/news/speech/spe_achievable.htm (accessed 22 November 1999).

34 Chris Fay (chairman and chief executive, Shell UK Ltd), 'Achievable targets needed', speech at the CBI Panel Debate, 'Climate Change – a Taxing Business?', at the CBI National Conference, Birmingham,

10 November 1997. Source: www.shell.co.uk/news/speech/
spe_achievable.htm (accessed 22 November 1999).

35 Personal communication with Gerry Matthews, Shell International, Washington, DC, March 2000.

36 Personal communication with Gerry Matthews, Shell International, Washington DC, March 2000.

37 Thus, ExxonMobil's net income in 2000 was roughly 13 times larger than Statoil's. Shell's net income in the same year was roughly 10 times larger than Statoil's.

38 Source: www.statoil.com (see also Norwegian White Paper no. 36, December 2000).

39 Source: www.statoil.com.

40 'Statoil and climate policy'. Source: www.statoil.com.

41 'Statoil and climate policy'. Source: www.statoil.com.

42 Statoil press release, 26 May 2000. Source: www.statoil.com.

43 Statoil press release, 26 May 2000. Source: www.statoil.com.

44 *Stavanger Aftenblad*, 28 January 2000.

4
The Corporate Actor model

The previous chapter demonstrated the striking differences in the climate strategies of ExxonMobil, the Shell Group and Statoil. While ExxonMobil has adopted a reactive strategy, Shell has chosen a proactive response, and Statoil has adopted a strategy representing a hybrid between these two positions. In this chapter we explore the explanatory power of the approach we have labelled the Corporate Actor (CA) model.

To recapitulate our discussion from chapter 2, the CA model suggests that differences in the companies' climate strategy choice are explained by differences in the companies themselves. The business environmental management literature suggests a host of company-specific factors that may have an impact on strategy choice in relation to an issue such as climate change. We have chosen to focus on three main factors: (1) the environmental risk associated with current and future corporate operations; (2) the environmental reputation of the company; and (3) the company's capacity for organisational learning. We assume that companies with low environmental risk, experience with negative public scrutiny, and high capacity for organisational learning (conditioned by other factors) are more likely to adopt a proactive climate strategy than are companies with high environmental risk, no experience with negative public scrutiny, and low capacity for organisational learning.

The company-specific factors we have chosen to analyse in depth in this study are selected from a long list of factors suggested to have an impact on corporate strategy choice in an issue area such as climate change. While we assume that the factors we have chosen to focus on are the most important for explaining differences in corporate strategy choice in this case, there may nevertheless be other company-specific factors not analysed in depth here that may modify our conclusions. Thus, alternative company-specific factors are briefly discussed in the last section of the chapter.

Environmental risk

One factor assumed to have an impact on a company's choice of environmental strategy is the environmental risk associated with its activities. The risk of climate change is linked to fossil-fuel combustion and increases with the carbon intensity of the fuel. Thus, the environmental risk of a company mainly engaged in coal activities is higher than that of a company mainly engaged in oil and natural gas activities. It is thus reasonable to assume that the more carbon intensive the fossil-fuel portfolio of the companies is, the higher is their risk of being subjected to more stringent regulation, and the more likely they are to resist such policies and adopt a reactive strategy. Most multinational petroleum companies have their main activities in coal, oil and gas. The argument thus relates to differences in the *relative* importance of coal and oil versus natural gas in the companies' portfolio of fossil-fuel activities: according to this logic, oil companies with relatively more emphasis on coal and oil are more likely to adopt a reactive climate strategy than are companies with a larger relative stress on natural gas.

The environmental risk associated with the oil industry's operations in relation to the climate problem is thus analysed in terms of each corporation's main areas of activity – now and in the future (reserves). This gives us an indication of the companies' relative emphasis on coal and oil versus gas in their fossil-fuel portfolio. This approach also gives an indication of the extent to which the companies operate in the same market, and thus whether they are confronted by the same challenges and opportunities in their choice of strategy to deal with this problem.

Shell and ExxonMobil have very similar portfolios in the sense that their key business areas are oil and gas exploration and production, and chemicals manufacturing. Production figures for the two companies show close similarities. ExxonMobil's 2000 production included 2.6 million barrels daily of crude oil and natural gas liquids (NGL) and 10.3 billion cubic feet daily of natural gas available for sale (ExxonMobil, 2000a). The corresponding figures for Shell include 2.3 million barrels daily of crude oil and NGL and 8.2 billion cubic feet daily of natural gas available for sale (Shell, 2000b). Until 1999, both companies produced coal. In 1999, Shell's production of coal rose 20 per cent relative to 1998, to 17.1 million tonnes (Shell, 1999). ExxonMobil's coal production in 1999 rose 12.6 per cent relative to 1998, to 16.9 million tonnes (ExxonMobil, 1999a). Shell divested its coal assets in 2000. ExxonMobil's coal production in 2000 was 16.6 million tonnes (ExxonMobil, 2000a).

Statoil's business portfolio resembles that of the other two, although on a smaller scale. Statoil's 2000 production included 509,000 barrels daily of crude oil and NGL, and 773.5 million cubic feet daily of natural gas (Statoil, 2000).[1] Statoil, however, manages the Norwegian SDFI on the Norwegian continental shelf in addition to its own equity interests. If we add the production numbers for the SDFI to Statoil's own production, the company joins the league of the world's largest oil companies.[2] Together, Statoil and the SDFI produced 1.8 million barrels daily of crude oil and NGL and 3 billion cubic feet daily of natural gas. Statoil, however, does not produce coal.

Both ExxonMobil and Shell have major oil and gas reserves. In 2000, ExxonMobil's worldwide total reserves were 11.6 billion barrels of crude oil and NGL and 56 thousand billion cubic feet of natural gas. The reserve replacement ratio in 2000 was 110 per cent.[3] The corresponding figures for Shell are 8.8 billion barrels of crude oil and NGL and 51 thousand billion cubic feet of natural gas. The reserve replacement ratio in 2000 was 105 per cent (ExxonMobil, 2000a; Shell, 2000b) Statoil's reserves are significantly less, at 1.5 billion barrels of oil and NGL and 8.2 thousand billion cubic feet of natural gas. The replacement ratio of Statoil's reserves in 2000 was 86 per cent (Statoil, 2000). Together with the SDFI reserves, however, Statoil manages reserves that are of a comparable size, particularly with regard to natural gas: 4.8 billion barrels of oil and NGL and 38 thousand billion cubic feet of natural gas.

Transformed into oil equivalent barrels, the figures show that ExxonMobil is more carbon intensive than Shell and Statoil. The carbon intensity of ExxonMobil is further increased by its coal reserves. Since Shell divested its coal assets in 2000, ExxonMobil has been the only company of the three that has operations in coal. It is important to emphasise, however, that, as discussed above, we found reason to regard Shell's divestment of its coal activities as an *effect* rather than a *cause* of the company's climate strategy (see chapter 3). When Shell adopted its proactive climate strategy, both Shell and ExxonMobil produced coal and both controlled large amounts of coal reserves. The difference between Shell and ExxonMobil in carbon intensity, therefore, lies in ExxonMobil's relatively larger reserves of oil than gas. Shell and Statoil (without the SDFI) have approximately equal reserves of oil and gas, while the Statoil/SDFI portfolio is the least carbon intensive, with significantly larger reserves of gas than oil (see table. 4.1).

Table 4.1 *Oil and gas reserves in 2000: ExxonMobil, Shell and Statoil*

Company	Crude oil and NGL (billion barrels)[a]	Natural gas (billion ft)[3]	Natural gas (billion barrels oil equivalent)
ExxonMobil	11.6	56,000	9.7
Shell	8.8	51,000	8.8
Statoil	1.5	8,200	1.4
Statoil/SDFI	4.8	38,000	6.5

Note: [a]Barrels of crude oil and NGL are equal to barrels of o.e.

The main markets for both ExxonMobil and Shell are the US and Europe. There are differences between the two companies, however, in the *relative* importance of the European and the US markets. While the European market is more important to Shell, the US is more so to ExxonMobil. Shell's total petroleum product sales in Europe were more than double their sales in the US in 1998 (Shell, 1998a). Similarly, ExxonMobil sold significantly more petrol in the US than in Europe, and about twice as much in the US as Shell (ExxonMobil, 1998). The main markets for Statoil's crude oil trading are north-western Europe, the US and Canada, while the company's retail activity is largely concentrated in the Nordic region and Europe, with a 25 per cent share of the Scandinavian market (Statoil, 2000).

The overall picture of the companies, therefore, is that in terms of core business areas, exploration and production volume, and resource reserves they are very similar and highly comparable. In contrast to environmental regulations to abate air and water pollution from the oil industry, a vigorous climate policy threatens the core business of oil companies. Thus, as oil companies, the three all face a considerable environmental risk in climate policy as compared to other industries. The most striking observation, however, is perhaps that none of the three companies has any incentives to adopt a proactive strategy, although there are some important relative differences: ExxonMobil is somewhat more carbon intensive than the other two, which indicates a higher score on environmental risk than Shell and Statoil, and hence a somewhat stronger incentive for choosing a reactive strategy. This observation is consistent with the CA model, since ExxonMobil has chosen the most reactive strategy of the three. On the other hand, the low carbon intensity of the Statoil/SDFI portfolio – which would pull in the direction of a more proactive strategy choice according to the CA model – does not match well with the strategy actually chosen by Statoil. This finding, however, is modified by the relationship between Statoil and the SDFI and the fact that the two do not constitute one company. Nevertheless, the CA model is not capable of distinguishing between Shell and Statoil: because of Shell's coal assets at the time of its strategy choice, Statoil is less carbon intensive than Shell even without the addition of the SDFI portfolio and would thus, according to the CA model, have higher incentives for choosing a proactive strategy. Thus, there is a mismatch between the strategies actually adopted by these companies and those predicted by the CA model.

Environmental reputation

A key factor in a company's perception of the environmental risk and opportunities associated with its business operations is its experience with public exposure and criticism in relation to environmental and political incidents. A company's environmental reputation can thus be assumed to have an impact on its choice of climate strategy. This reputation may affect the company's choice of climate strategy in the sense that companies with experience of strong negative public scrutiny will seek to avoid such scrutiny and are thus more likely to respond to an enhanced public concern for climate change by adopting a proactive climate strategy. Thus, it is our assumption that a negative environmental reputation induces corporations to choose a proactive rather than a reactive climate strategy.

ExxonMobil, Shell and Statoil all have experience of public exposure and criticism in relation to environmental and political incidents related to their operations, although there are differences in the scale and intensity of the attention to which they have been subjected. The legal process against Exxon is still not concluded, more than a decade after the 240,000-barrel *Exxon Valdez* oil spill in Prince William Sound in Alaska in 1989. The spill was the largest in US history – more than 11 million gallons of crude oil – and threatened the delicate food chain that supports Prince William Sound's commercial fishing industry: more than 11,000 people and businesses were affected to the extent that they have received compensation. Exxon has spent US$2.2 billion on the clean-up, which continued until 1992, in addition to large payments in compensation and damage claims, totalling US$3.5 billion that the corporation has spent on the spill.[4] In 1994, a jury ordered Exxon to pay US$287 million in compensatory damages plus US$5 billion in punitive damages for behaviour that led to the 1989 oil spill.[5] While Exxon did not dispute the ruling on compensatory damages, it has vigorously objected to the huge punitive award. In 2001, the Ninth Circuit Court of Appeals overturned the punitive damages award against Exxon and ordered a district court to set a new, lower amount. The case has now been sent back to the federal court in Anchorage, Alaska.[6]

The *Exxon Valdez* accident has meant more than economic losses for the company. Exxon's environmental reputation was at stake and the company lost credibility. Exxon is still criticised for arrogant behaviour. ExxonMobil officially regrets the incident and argues that the company has paid an enormous price, that it has made significant technical and procedural changes to prevent a similar spill, and that the environment in Prince William Sound is healthy, robust and thriving. The victims, however, maintain that Exxon has made profits on the delayed award, manipulated scientific information and presented lies since 1989. For example, the oil spill Trustee Council claims that only two species have recovered fully from the impact of the spill.[7]

The Shell Group has also been severely exposed to public indignation as a result of several incidents of both an environmental and a political nature. In the anti-apartheid campaigns of the 1980s, Shell was subjected to a massive, worldwide consumer boycott for its operations and activities in South Africa. Two more recent incidents both took place in 1995: first, when Shell was indirectly linked to serious violations of human rights in Nigeria; and second, in relation to its attempted deep-sea disposal of the redundant North Sea installation, Brent Spar. In this context, however, it is interesting to note Grolin's observation that corporate legitimacy depends on perception or assumption rather than matters of fact, meaning that an organisation may violate social norms and yet retain its legitimacy as long as the violation goes unnoticed. Grolin's pertinent example is that Exxon was co-owner of the Brent Spar, but 'ducked its head and left it to Shell to take the heat' while the Brent Spar conflict lasted (Grolin, 1998: 216). It should be noted, however, that it was in Shell's capacity as operator that Greenpeace targeted Shell alone on the Brent Spar issue.[8]

While Greenpeace initially targeted only Shell on the climate issue, the recent change in US climate policy by the Bush administration has also brought about a renewed public interest in ExxonMobil's actions and inaction in this area. In April 2001, Greenpeace International announced its climate campaign against US oil companies – including ExxonMobil – with an aim 'to hurt their markets outside the United States until they withdraw their support for the Bush administration's rejection of ... the Kyoto Protocol' (Greenpeace International, 2001). In May 2001, Greenpeace, in alliance with Friends of the Earth and People and Planet, launched its 'Stop Esso' campaign – a UK boycott of Esso, the European counterpart of ExxonMobil – over its support for the US withdrawal from the Kyoto agreement.[9] In July, Greenpeace reported that the campaign was gradually gaining momentum, had spread to Norway and Finland and was beginning to establish itself in Germany and the Netherlands.[10] ExxonMobil denies allegations made by campaigners that it does not invest in renewables and claims to have made such investments to the order of US$500 million.[11] It turns out, however, that this investment was made in the 1980s, independent of the climate-change problem. Moreover, this investment is regarded by the corporation as a failure – a failure that also constitutes one of the main reasons given by corporation officials for the stance it has adopted towards renewables since the 1990s[12] (see also chapter 3). In November 2001, more than 300 UK Esso filling stations were targeted by around 3,000 protesters in a bid to urge motorists to boycott the company because of its stance on the climate issue.[13]

When the campaign was launched, ExxonMobil officials declined to comment on the UK boycott's impact on sales. Analysts maintained, however, that the campaign would be damaging to any oil company operating in the notoriously competitive UK fuel market, where retail margins have been 'wafer thin' for some time and oil companies took some time to benefit from an upturn in profits after protests in 2000 forced retailers to hold their prices down.[14] In July 2002, a survey showed that the campaign indeed did have an effect on British motorists' choice of petrol retailers[15] (see chapter 5).

While Statoil has also experienced negative publicity over environmental and political incidents, the company has only to a very small extent been subjected to *international* public scrutiny: attention has usually been limited to Norway or the Nordic countries. Given that these are the company's main markets, however, the impact may be equally serious. Also, while Statoil has received public attention and criticism, it has not been subjected to direct and explicit boycott campaigns either at home or abroad. In the mid-1990s, Statoil, like Shell, was criticised for its activities in Nigeria during a period when serious violations of basic human rights were indirectly linked to the oil companies with operations there. On this occasion, Statoil was not explicitly targeted internationally because Shell – a bigger and thus more 'target-prone' corporation – was. Greenpeace's current campaign to keep the oil industry out of the Arctic Ocean also implies a serious incident of negative attention for Statoil. After having campaigned against BP's Northstar offshore oil project in the Beaufort Sea since 1997, Greenpeace opened up what was referred to as 'a second front' against oil activities in the Arctic in March 2000, when Greenpeace's Nordic office launched its campaign against Statoil's development of new fields in the Arctic Barents Sea because of the potential harm to the region's fragile ecological system.[16] On a national scale, the most controversial issue for

Statoil has been its involvement in Naturkraft (one-third ownership) and its plans to build Norway's first gas-fired power plants.
These plans have been criticised by the entire community of environmental NGOs in Norway, including Greenpeace, Friends of
the Earth and the Norwegian NGO Bellona.

Exxon's response to the *Exxon Valdez* accident has been
directed towards the prevention of similar accidents: modifica-
tions of tanker routines, the institution of drug and alcohol
testing programmes for employees in sensitive positions, more
rigorous training programmes, more extensive periodic assessments of vessels and facilities, and so on.[17] Shell, on the other
hand, initiated a major reorganization process of the whole
corporation in 1995, partly in response to the public scrutiny it
experienced.[18] The incidents related to Shell's activities thus seem
to be perceived, by Shell, as a real threat to its corporate legitimacy and credibility. In Statoil, similar organisational responses
to negative public scrutiny cannot be identified. The main motivation for the restructuring process initiated in the latter half of
the 1990s was to strengthen earnings and competitiveness.
During this period, however, Statoil also instituted a more
systematic approach to its social responsibility.

All the three companies, therefore, have experienced negative
public scrutiny in relation to environmental and political incidents, although of a different nature and on a different scale.
Statoil, being the smallest of the three companies, has to a lesser
extent than ExxonMobil and Shell been exposed to negative
attention at the international level. On the other hand, Statoil has
been exposed to public criticism in its main markets. And
Greenpeace is presently targeting Statoil together with BP for its
activities and plans in the Arctic region. On this basis, we can
conclude that all three companies have been exposed to public
scrutiny to the extent that they have incentives to adopt a proactive environmental strategy in order to avoid loss of reputation.
The high degree of similarity between the companies on this
dimension is more striking than the differences. This observation
is not in line with the CA model in view of the actual climate
strategies adopted by these companies. Differences in public
scrutiny apparently do not account for the observed differences in
the climate strategies of the companies.

Capacity for organisational learning

The environmental risk associated with a company's portfolio of business areas and the environmental reputation a company enjoys represent external sources of impact on a company's climate strategy choice. Another set of factors that may have impact on strategy choice stems from internal sources. In this analysis we focus on one such factor: a company's capacity for organisational learning.

As discussed in chapter 2, organisational learning basically concerns two main dimensions. The first is a company's capacity to capture signals of trends and trace changes in areas of relevance to its business, particularly in terms of the extent to which it has institutionalised a systematic monitoring of future trends. The second is a company's capacity to make use of and internalise the knowledge generated through monitoring mechanisms, which is particularly linked to a company's organisational structure.

Monitoring of trends

Since 1971, *Shell* has explicitly addressed issues of uncertainty in its strategy formulation through scenario planning. The scenario approach developed by Shell is based on the understanding that 'the only competitive advantage the company of the future will have is its managers' ability to learn faster than their competitors' (de Geus, cited in Neale, 1997:96). According to Shell, scenarios 'assist in the understanding of complex situations, providing a useful tool for organisational learning' (Shell, 1998c: 1). Shell's scenario approach constitutes a central element in its formulation of a climate strategy. Shell anticipates a future in which low-carbon and renewable energy sources may cover up to as much as 50 per cent of world energy demands by 2050.

In its current scenarios, Shell explicitly addresses corporate challenges that arise from the effects of processes of change that operate at two levels: (1) that of markets, financial systems, governments and other wide-reaching institutions, particularly in terms of globalisation, liberalisation and technological innovation; and (2) that of people, particularly in terms of changes related to education, wealth and choice. With regard to the latter source of change, interestingly Shell maintains that 'in developed countries, wealthier people express a greater willingness to pay for goods and services that are "green" or that are the products of socially responsible companies' and explicitly addresses the question of 'what will be the effects of more people becoming wealthy enough to make such choices?' (Shell, 1998c: 8). Thus, in its scenario planning and development Shell encompasses a much wider spectrum of uncertainties than the more traditional focus on issues such as oil price, political and financial trends and the post-Cold War world order (Neale, 1997).

Against this background, Shell has developed two scenarios: *The New Game* and *People Power*. In *The New Game*, new institutions emerge and old ones are reconstructed to cope with processes of globalisation, liberalisation and technological innovation. The global scope of *The New Game* implies that success comes to relatively few players in the field: 'The good players reap extremely large rewards, while those who play poorly struggle simply to survive' (Shell, 1998c: 11). Faster-paced processes imply that learning constitutes a necessity: 'Businesses and institutions do best in The New Game when they are designed as learning systems, continually reinventing themselves' (Shell, 1998c: 12). In this scenario, Kyoto works. Targets are achieved through the trading of carbon emission permits, and this new market serves to drive out subsidies and increase reliance on rule-based systems. It also drives coal out of the energy mix in OECD countries (Shell, 1998c: 15).

In the other scenario – *People Power* – the main characteristic feature is that 'large numbers of people across the globe are free to express their own values and often do so in unpredictable, unstructured, and spontaneous ways', combined with increases in wealth, choice and education (Shell, 1998c: 19). While this gives rise to an unprecedented level of diversity, which also leads to a fragmentation of political parties and an undermining of institution's ability to build consensus, it also facilitates like-minded peoples' ability to express their values through a variety of global associations. In *People Power*, Kyoto does not work: it is ratified by the EU, but not by the US. This leads to public outrage: 'Angry about local pollution, congestion, and health issues, protesters target oil, coal, and car companies through increasingly effective NGOs and individual action ... Corporations, under intense media scrutiny, are held to higher standards of social accountability – People Power in action' (Shell, 1998c: 23).

Both of these scenarios imply a future where the Kyoto Protocol is ratified – either partially (by the EU in *People Power*) or fully (by the OECD in *The New Game*). Both scenarios imply a stronger demand for 'green consciousness' in large corporations. In *The New Game*, this demand is voiced through new institutions and a higher willingness to pay, whereas in *People Power*, it is voiced through more effective, global interest and value associations and NGOs. Thus, in its primary tool for organisational learning, Shell foresees a future in which global corporations operate within a framework of international or regional governmental regulations on energy and where they are forced – through an increased public demand – to develop a portfolio of cleaner fuels. In this regard, it is not only the existence of a tool for organisational learning itself that distinguishes Shell from the other companies. The manner in which Shell employs this tool also gives an indication of the widely different worldviews and perceptions that characterise the corporations.

ExxonMobil does not approach the task of monitoring future trends in the same systematic manner as Shell. We have only been able to identify one Exxon scenario – a short-term one covering the period 1990–2010 (Marriott, 1991; Estrada et al., 1997). This was presented at the Global Climate Change Symposium organised by the International Petroleum Industry Environmental Conservation Association (IPIECA) in Rome in April 1991. Based on trend analyses of economic growth, energy intensity, fuel availability, patterns of world energy use and power generation, the 1991 Exxon scenario generally concludes that 'the energy economy of 2010 will in all probability remain largely built on the use of fossil fuels' (Marriott, 1991: 121). In this scenario, no technology breakthroughs for renewables are anticipated, although it is stated that 'some may reach the early stages of commercial application by the end of the period' (Marriott, 1991: 121). Because of large indigenous supplies, the Exxon scenario expects a growth in the use of coal for power generation, at least in absolute terms, particularly in developing countries like China and India, but also in the United States. This is an interesting contrast to one of the Shell scenarios, which anticipates that coal is driven out of the energy mix in OECD countries, although the widely differing time perspectives in the two scenarios should be noted. It is also interesting to note that in contrast to the Shell scenarios, Exxon's scenario does not take government (environmental) policies explicitly into account. While it is acknowledged that 'factors such as government policy, pricing, technology development, environmental standards, etc., are important determinants of future energy demand', the effects of these factors are reflected only in terms of changes in economic activity and energy intensity (Marriott, 1991: 123). Also in sharp contrast to the Shell scenarios, the Exxon scenario does not consider changes in the public's demand for 'clean energy'. Consequently, Exxon also does not consider the enhanced risk for shaming campaigns that Shell foresees for industries like the oil industry as a result of this public concern. Finally, the Exxon scenario emphasises that the continuing fossil-fuel-based energy mix implies that 'reductions in carbon emissions towards world stablization cannot be achieved without a significant economic penalty and/or by imposing severe restrictions on energy use detrimental to individual lifestyles' (Marriott, 1991: 133). In sum, therefore, there are sharp contrasts between Shell and ExxonMobil with regard to the extent to which systematic monitoring is an institutionalised task within the company and particularly with regard to the two companies' perceptions of the future.

Statoil also does not approach the task of monitoring future trends in the same broad-based and systematic manner as Shell, although the company uses the scenario method to project future long-term trends in oil and gas markets. In 1992/1993 Statoil developed an environmentally aware energy-driven market scenario, which foresees a future where an increased environmental awareness among consumers 'can trigger a trend characterised by diminishing use of fossil fuels'.[19] This may lead to a reduction in the demand for oil, resulting in a 'substantial overcapacity' in the industry, investments in new fields, and a scaling down of oil activity (Estrada et al., 1997: 143). In this scenario, natural gas is seen as the 'bridge' to alternative energy carriers. The time-span suggested for this transitional period is 50–60 years. For Statoil, therefore, more natural gas, not renewables, is the short-term implication.

This brief discussion of the monitoring activities in the three companies shows, first, that there are significant differences in the extent to which systematic monitoring is institutionalised within the company. Having implemented the scenario approach over three decades, Shell has the most systematic and broad-based monitoring system of the three. While both ExxonMobil and Statoil also employ the scenario method, it is less systematic. These companies, moreover, are less open to the external public with regard to the specifics of their perspectives of the future. Second, there is a sharp difference particularly between the Exxon and Shell scenarios in *what* is monitored. Exxon presents a more 'traditional' scenario, whereas Shell seems to have much more flexibility as to which indicators and trends that are monitored. For instance, while future government regulations and public demands constitute central elements in Shell's perspectives of future risks and challenges, these factors are almost completely absent from the Exxon scenario. There is also a difference between Statoil and Shell in this regard, although it is smaller, particularly given Statoil's view of changes in consumer behaviour on the basis of prospects for a growing public environmental consciousness. Our analysis thus shows that there are persistent differences between the three companies in the extent to which they have institutionalised a systematic monitoring of future trends, as well as in which trends are monitored, and hence in the companies' perspective on future risks and challenges.

Organisational structure
As discussed in chapter 2, organisational structure is seen as one key factor determining a company's capacity to make use of the knowledge generated through monitoring activities, with a rough distinction running between centralised and decentralised companies. It can be argued that a centralised company is better equipped for internal communication and coordination, and thus has a larger capacity to make use of information generated through monitoring. A decentralised company, on the other hand, would be less capable of communicating trend shifts from one part of the organisation to another, and would thus also be less capable of internalising this kind of information.

All of the three companies have undergone more or less substantial organisational changes during the 1990s. The reorganisation of Shell probably represents the most significant. Until Shell initiated its reorganisation process in 1995, the company was characterised by a strongly decentralised structure. From 1959 to 1995, the company was organised according to the

'McKinsey-derived matrix structure', which was unusually complex:

> There was no corporate headquarters, and the service companies, based in London and The Hague, were organised in a 3-way matrix, with functional areas disseminating professional and technical advice throughout the group, business sectors giving strategic advice to the operating companies and regions scrutinising capital expenditure plans and appraising operating companies' performance on behalf of the shareholders. Most decision making was concentrated at the 100-odd local operating companies. (Neale, 1997: 96)

The complexity of the structure was further enhanced by Shell's emphasis on reaching decisions by consensus, which implied that decision-making in Shell 'involved an unusually high level of internal discussion' (Neale, 1997: 96).

Although there were certain institutionalised arenas to coordinate the strategies and activities of the company, this structure was not very well suited for corporate control and efficient communication and decision-making. Thus, while all activities carried out by any of Shell's many diverse branches and units were accredited to Shell's brand name, the two head offices of Shell – Shell UK and Shell Netherlands – did not have effective instruments to control them. Shell Oil in the US, for instance, constituted an almost independent unit, over which Shell's main offices in Europe had little or no control. It seems quite clear, moreover, that the reorganisation process was, at least partly, initiated as a response to the massive negative public scrutiny the company experienced from the mid-1980s.[20] According to Neale, the new structure 're-creates a single, functional, line of command, and is intended to deliver "greater clarity of roles and responsibilities"' (1997: 101, citation by Neale from Shell, 1996). Currently Shell is organised in five core business areas; exploration and production, oil production, chemicals, gas and power, and renewables (see figure 4.1).

Exxon has also undergone an organisational change, not least in terms of its merger with Mobil in 1999. Before the merger, Exxon was a highly centralised organisation with six major divisions.[21] The centralised structure implied that corporate headquarters played a major role in the decision-making process of the company, particularly with regard to investment decisions. With

Figure 4.1 *Company structure: the Shell Group*

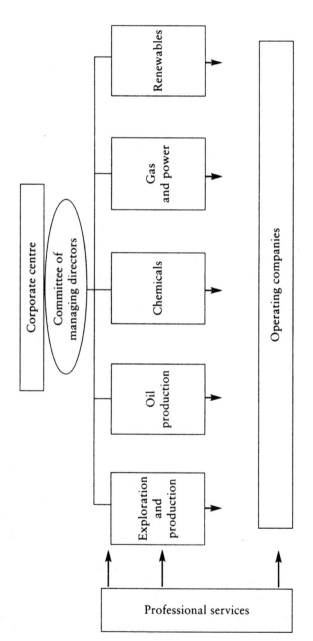

Source: Shell (2000c).

its merger with Mobil, the company moved from a multifunctional, geographically based regional organisation to 11 global functional businesses organised in four core business areas: upstream, downstream, chemical, and power, coal and minerals. Thus, ExxonMobil has moved in the opposite direction to Shell, with the merger bringing about a more decentralised structure. Headquarters nevertheless still play a central role in the decision-making process of the corporation (see figure 4.2).

The reorganisation of Statoil is particularly linked to two events: first, the company was reorganised from fifteen to five business areas in 1999, and second, it was partly privatised in 2001, when it was listed on the Oslo and New York stock exchanges. The five core business units of Statoil are exploration and production Norway, international exploration and production, European gas, manufacturing and marketing, and the SDFI. Statoil represents a relatively centralised organisation (see figure 4.3).

With their current organisational structures, therefore, the three companies are very similar. They are all organised in relatively centralised structures with corporate headquarters that govern all parts of the organisation according to the same principles and that ensure communication across all divisions. For ExxonMobil and Shell in particular, this vertical structure is supplemented with cross-cutting units that provide various types of services across the organisation.

Capacity for organisational learning: in sum   Our discussion shows that while there are differences in the learning capacity of the three companies, particularly with regard to the degree to which they have institutionalised monitoring systems, the similarities are nevertheless dominant. With its long tradition of using the scenario approach, Shell has a higher score on the monitoring dimension of organisational learning than the other two. On the other hand, Shell also, with its highly decentralised structure, is the company that until its reorganisation in 1995 had the lowest organisational capacity to make use of the knowledge generated through its monitoring system. In this perspective, Shell's reorganisation contributed to an enhancement of the company's learning capacity during the 1990s. While the other two companies have adopted less systematic and broad-based approaches to

Figure 4.2 *Company structure: ExxonMobil*

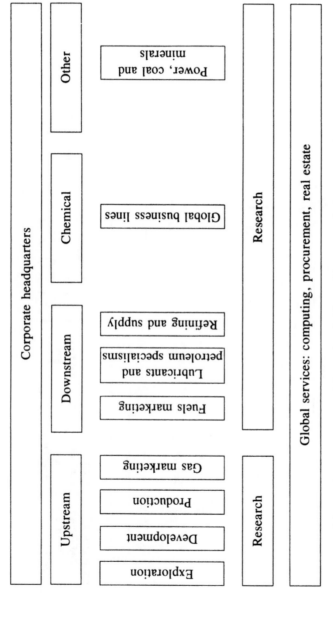

Source: ExxonMobil (1999b), www.exxon.mobil.com/shareholder_publications/c_fo_99/c_merger.html.

Figure 4.3 *Company structure: Statoil*

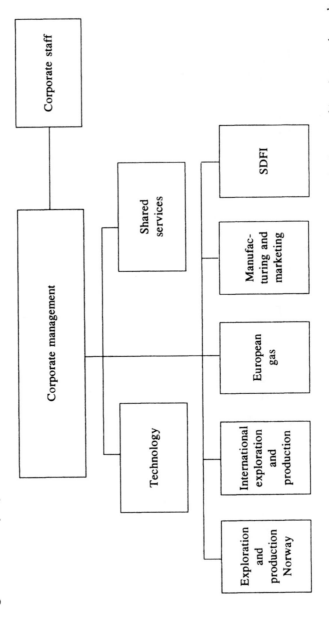

Source: Statoil press release, 'New structure: the Statoil group is restructuring its organisation to achieve improved results', 19 February 1999.

their monitoring of future trends than Shell, they nevertheless conduct trend analyses and they have an organisational structure conducive to making use of the knowledge generated. On this basis, we conclude that ExxonMobil and Statoil have a medium capacity for organisational learning while Shell has a high capacity. The CA model predicts that the higher the learning capacity, the more likely the company is to choose a proactive climate strategy (conditional on other factors). Thus, the slight difference we have identified in the companies' capacities to learn may have contributed to the observed difference in their choice of climate strategies, although the difference is too small to provide a satisfactory explanation. Moreover, there is a mismatch between the similarities in the learning capacity of ExxonMobil and Statoil and the difference in their choice of climate strategy.

Shell has undergone the most substantial organisational change, and this has enhanced the company's learning capacity. To what extent can this change explain the company's shift in climate strategy? Shell officials themselves maintain that the organisational change initiated in 1995, first, was a response to the company's inability to deal with (new) social expectations, particularly illustrated by the shaming campaigns the company was exposed to in the mid-1990s. Second, they maintain that the reorganisation also brought about new perspectives on the climate issue, and a new understanding of risk within the company.[22] The time lag between the initiation of the reorganisation (1995) and the shift in strategy choice (1997/1998) indicates that there is no direct link between these two events. Moreover, it is important to recall that Shell as late as 1996 was involved in an effort to prevent the establishment of a regulatory regime for GHG emissions reductions, which indicates quite clearly that the company's revision of its climate strategy had not even started at this point in time – one year after the initiation of the reorganisation (Leggett, 1999; see chapter 3). We thus conclude that while Shell's reorganisation may have contributed to its shift in climate strategy, this factor does not provide a satisfactory explanation of the shift.

Perhaps the most interesting finding in this exploration into the three companies' capacity to learn is the striking differences we have identified in which trends the companies monitor. The difference between ExxonMobil and Shell is particularly notable,

and indicates that these companies have strongly diverging
perceptions of which trends and trend changes affect their core
businesses and hence need to be monitored. While the Shell
scenarios indicate a strong focus on shifts in policy and shifts in
public sentiments and demands, these factors are almost
completely absent from the Exxon scenario. This difference could
have been explained, for instance, by differences in the companies' experience with negative public scrutiny. As our analysis
also shows, however, all the three companies have roughly similar
experiences on this dimension. Thus, we believe that the differences in the companies' perceptions of which trends pose risks
and challenges to their core interests and hence need to be monitored reflect actual differences in the (home-base) contexts in
which the companies operate. This point will be further explored
in chapter 5.

The explanatory power of the CA model

To recapitulate, we assumed that the companies with low environmental risk, experience with negative public scrutiny and a high
capacity for organisational learning are more likely to adopt a
proactive strategy than companies with high environmental risk, no
experience with negative public scrutiny and a low capacity for
organisational learning. The main results are presented in table 4.2.

Table 4.2   *The CA model: expected versus actual strategies in relative
terms*

Company	Risk	Negative public scrutiny	Learning capacity	Expected strategy[a]	Actual strategy
ExxonMobil	High	High	Medium	Intermediate(−)	Reactive
Shell	Medium	High	High	Intermediate(+)	Proactive
Statoil	Medium	Medium	Medium	Intermediate	Intermediate

[a] In the absence of 'pure' cases, the (−) and (+) signs are added to indicate the directions in which risk and learning are pulling, i.e. towards a reactive strategy for
ExxonMobil and towards a proactive one for Shell.

From table 4.2 we see that the propositions derived from the
CA model gain limited empirical support judged on the merits of
pattern matching. While two cases do not match with expecta-

tions derived from the CA model, we have one case that matches. While the climate strategies adopted by the three companies vary significantly, the companies actually show more similarities than differences with regard to environmental risk, exposure to negative public scrutiny and learning capacity. Perhaps more importantly, the CA model largely fails to explain *changes* in corporate strategies. In particular, while the shift in Shell's learning capacity with its reorganisation in 1995 may have contributed to the company's turnabout on the climate issue (1997/1998), the time lag between the two events is considered too large to indicate any direct link. More importantly, Shell was involved in a concerted attempt to prevent the development of a climate regime in 1996, one year after the reorganisation was initiated. Moreover, the *Exxon Valdez* accident in 1989 was disastrous, and is still negative for ExxonMobil's environmental reputation, but the accident did not lead to any change in the company's orientation towards climate change.

On the other hand, the differences observed particularly between Shell and Exxon pull in the directions predicted by the CA model. We have identified differences on two dimensions. First, ExxonMobil is more carbon intensive than Shell and Statoil. ExxonMobil's reserves of oil are larger than its reserves of natural gas. In addition, the company has operations in coal, which is the most carbon intensive of the fossil fuels. As discussed above, however, the difference with regard to coal should not be ascribed any weight in explanations of Shell's climate strategy, since there are reasons to view the company's divestment of its coal assets as an effect of its climate strategy rather than as a cause of it. It should also be noted that while ExxonMobil is the more carbon intensive company, Shell and Statoil are nevertheless also among the world's largest companies in the petroleum business. There is therefore a significant environmental risk associated with their operations as well.

Second, with its systematic approach to monitoring environmental trends, Shell has a higher capacity for organisational learning than the other two companies. It is also interesting to note that this difference goes beyond the organisational device itself in its indication of strongly differing worldviews, perceptions and anticipations of the future according to which the companies develop their strategic plans. While ExxonMobil puts a lot of effort and money into lobbying against governmental GHG regulations, both of Shell's current scenarios include a ratification of the Kyoto agreement in some form or other (with or without the US).

These differences correspond well with our initial proposition about the relationship between these company-specific factors and strategy choice: high carbon intensity and hence high environmental risk was assumed to increase the likelihood of a reactive strategy, while high capacity for organisational learning was assumed to increase the likelihood of a proactive strategy.

Along the other dimensions of the model, however, the match between the companies' scores on the indicators and their choices of climate strategies is poor. First, ExxonMobil and Shell have both experienced serious incidents of negative public scrutiny with a corresponding negative impact on their environmental reputation. Statoil also has this kind of experience, although on a smaller scale. Thus, in terms of this dimension, all the companies appear disposed towards a proactive strategy. Second, while Shell has a higher capacity for organisational learning due to its systematic approach to monitoring systems within its organisation, the three companies all have a high capacity for organisational learning in terms of their organisational structures. Since the mid-1990s, they have all operated within the frameworks of centralised structures that are well equipped for efficient internal communication and coordination.

In sum, we may conclude that while differences in company-specific factors, notably environmental risk and capacity for organisational learning, are pulling in the direction of the observed differences in climate strategy choice, the overall explanatory power of the CA model seems to be rather weak. However, the mismatch observed between expectations and actual strategies may be due to a too narrow specification of the CA model. As noted in chapter 2, a number of company-specific factors other than those included in the CA model have been identified as important in the literature. Below, we will explore some of these.

Moderating factors

Within the business environmental management literature, a host of factors other than those analysed here is often suggested as having an impact on a company's approach to (new) environmental problems. In this brief assessment of alternative factors that may modify our conclusions, we give attention to a set of company-internal factors, including corporate leadership, capital availability and human resource availability. In addition, we take a brief look at another factor that may be of some significance, namely ownership.

Corporate leadership

The impact of differences in leadership is intrinsically difficult to assess and 'measure' empirically for an external analyst and observer. There are indications, however, that there may be important differences between companies in leadership styles that also have implications for the companies' climate strategy choice. For instance, several references have been made to the unique role of Sir John Browne in directing BP towards a more proactive approach to climate change (see, for instance, Rowlands, 2000). Shell's chairman, Mark Moody-Stuart, who won *Tomorrow*'s 1999 Environmental Leadership Award, has also been attributed with some of the same qualities.[23] Lee Raymond, on the other hand, is often described as a more traditional, and conservative, corporate leader, notoriously sceptical about governmental intervention. He has reportedly described European suggestions that Americans should use smaller cars as neo-colonialism: 'In Europe you like to tell people what kind of cars they ought to use. Most Americans like to make that decision themselves – that's why they left [Europe].'[24]

Shell, moreover, has adopted a different management model to ExxonMobil's (Kolk and Levy, 2001). Shell operates with a committee of managing directors headed by a chairman instead of a CEO, with position changes every five years (Kolk and Levy, 2001; see also figure 4.2 above). This factor may work in several ways. First, a chairman has only a limited amount of time. Thus if he wants to make his mark on the company he needs to act swiftly and effectively. On the other hand, the frequency in leadership changes could also imply that the company generates a certain immunity to 'personal' leadership styles. It also raises the question of their long-term effects. Thus, while there are indications that the personal leadership styles of the big multinational oil companies have had an impact on these companies' strategic choice on the climate issue, there are also aspects that may work to prevent such effects.

Capital availability

According to available data, the companies are comparable in terms of net income and return on average capital employed. The year 2000 was an extremely good one for the oil industry. For ExxonMobil, it was also the first year of its merger with Mobil, and the company experienced a record net income of US$17.7 billion. Similarly, Shell had a record net income of US$12.7 billion. Both companies had a return on average capital employed of around 20 per cent. Statoil had a net income in 2000 of 11.3 billion Norwegian kr. (US$1.3 billion at an exchange rate of 8.85), which represents a return on average capital employed of 15.1 per cent. If we look at the numbers for 1999, which was a less exceptional year in terms of oil prices and which also excludes the effect of the Exxon–Mobil merger, the similarities persist: both Exxon and Shell had a net income approximating US$8 billion, with an approximate 10–12 per cent return on average capital employed (Shell, 2000a). Over the whole period between 1990 and 1999, however, Exxon had an approximately 3 per cent higher average return on capital than Shell (Kolk and Levy, 2001: 505).

Kolk and Levy (2001) argue that Exxon's higher profitability during the 1990s and the difficulties Shell and BP experienced in their financial situation during this period contributed to the observed differences in their market orientations (both BP and Shell have made investments in renewable energy sources while ExxonMobil has not) (see also chapter 3). This argument, however, may very well be turned the other way around: risky investments in renewable energy sources characterised by lower profitability (at least for the initial investment periods) may be more *unlikely* when companies experience a difficult financial situation. That is, in periods of low profitability, companies may be more reluctant to make new investments that do not give immediate payoff. Thus, the impact of differences in capital availability is also difficult to assess.

Human resource availability

Given that the companies operate in roughly the same markets, it is reasonable to assume that human resource availability is also roughly equal. A related point, however, is what types of human resources, in the form of different types of expertise, companies choose to have in-house, and what types of expertise they acquire, for instance, through alliances and cooperation. As pointed out in chapter 3, ExxonMobil is often characterised as a science- and technology-based corporation, in contrast to Shell and BP, which have internal technical and scientific expertise to a much lesser extent.[25] This is also evident from the organisational charts of the companies, where we see that ExxonMobil has cross-cutting research units that provide services to all divisions of the corporation (see figure 4.2). More specifically, with regard to climate change, ExxonMobil has had in-house climate expertise since the early 1980s, while Shell has relied on external sources of expertise to a much larger extent. This seems to have had an impact on the companies' responses to the scientific conclusions provided, for instance, by the IPCC – a point that will be further discussed in chapter 6.

Ownership

In chapter 2, we distinguished between three mechanisms through which ownership may have an impact on corporate climate strategy choice. The first concerns the impact of private versus state ownership. ExxonMobil and Shell are privately owned companies. Until recently, Statoil was fully owned by the Norwegian state. In chapter 2, we argued that while the effect of state ownership is difficult to predict, the close relationship between a company and its government owners suggests that national oil companies are more liable to choose a strategy that is in accordance with the position of their government owners on the climate issue. We analyse Norway's climate policy and its impact on Statoil's choice of climate strategy in chapter 5.

Second, another impact of ownership is seen in the extent to which shareholders exert pressure on companies to adopt a proactive climate strategy. When Texaco withdrew from the GCC in 2000, for instance, it was, in part in response to pressure from its shareholders.[26] Both BP and Shell have had similar experiences. On several occasions, for instance, shareholders have attempted to influence BP's activities in environmentally sensitive areas.[27] Greenpeace has also bought shares in Shell as a basis for similar attempts at coalition building with other shareholders to

affect the company's environmental strategies. After President
Bush's withdrawal from the Kyoto agreement, even ExxonMobil
has experienced similar actions from its shareholders. In May
2002, a religious shareholder group in ExxonMobil released a
report claiming that oil companies could find themselves facing
multi-billion dollar legal suits, similar to those facing tobacco
firms, if they ignore the potential consequences of global
warming.[28] At a subsequent annual meeting, the group succeeded
in mobilising support from 20.3 per cent of the shareholders in
favour of a resolution asking the corporation to adopt a renewable energy resources plan.[29]

Third, it has also been argued that differences in shareholder
patterns between Europe and the US have a strong impact on
corporate goals in the sense that managers of American corporations are fixated on more short-term financial performance (Pauly
and Reich, 1997). This would imply that Lee Raymond has a
more short-term perspective than Moody-Stuart, which may well
be true. Reportedly, Moody-Stuart believes that 'shareholders
generally, and people at Shell, are interested in the long term
rather than just the short term'.[30] However, it is extremely difficult to establish any causal links between such general observations and climate strategies.

Thus, even though Shell has experienced this type of pressure
from its shareholders over a longer period of time than
ExxonMobil, they have both been subjected to this type of tactic.
Also, ExxonMobil (together with Texaco) is the only company
that has been explicitly targeted on the climate issue. Thus, while
shareholder pressure may have had some impact and perhaps may
become even more important in the future, it is not likely to
explain the observed differences in climate strategies between
Shell and ExxonMobil.

Conclusion

None of these moderating factors seems to be capable of weakening our general conclusions to any significant extent. While most
of them may have some influence on the climate-strategy choice
of the companies, their impact is weak and/or its direction is
extremely difficult to specify. This brief assessment nevertheless
indicates that there are differences between the companies – for

instance in corporate culture and leadership style – that may affect the manners in which they deal with (new) environmental challenges. These differences, however, are subtle and their impact is difficult to measure in a systematic manner. Also, given that similar factors often pull in different directions, it seems reasonable to assume that these factors rarely have an independent and direct effect on climate strategy choice, but rather operate in combination with other factors. Thus, our main conclusion stands: while there are company-specific factors that contribute to explaining the observed differences in the climate strategies adopted by ExxonMobil, the Shell Group and Statoil, there is also a significant unexplained mismatch between the strategies chosen and those we would expect according to the CA model. Overall, the observed similarities in company features with relevance to their climate strategies are more striking than their differences. In addition, the CA perspective does not provide us with sufficiently good explanations of the changes observed in corporate strategies. The explanatory power of the CA model is thus judged to be limited.

Even though the scenarios produced by ExxonMobil, Shell and Statoil differ in their premises, purpose and time scales, it seems evident that these companies anticipate different futures: Shell sees a future based largely on renewables and the Kyoto Protocol; ExxonMobil has more faith in coal, oil and gas; and Statoil regards oil and natural gas as transitional energy carriers towards a non-fossil future. Moreover, these different scenarios are not systematically related to differences in previous corporate experience. For instance, both ExxonMobil and Shell lost money on renewables in the 1970s and 1980s, and they have both been exposed to public criticism during the 1990s. What they see depends on where they look. In the next chapter we explore explanations based on the assumption that the main sources of strategy choice lie in the political context characterising the home-base countries of the companies.

Notes

1 After Statoil was privatised in spring 2001, the company revised its production numbers for 2000 in accordance with its newly acquired shares of the SDFI. These numbers include 750,000 barrels daily of

oil and NGL and 1,412.8 million cubic feet daily of natural gas. Since this change only occurred in 2001, we use the numbers reported in their 2000 annual report.

2 It is important to note that Statoil and the SDFI are not one company, but Statoil itself used this argument in the debate preceding the partial privatisation of the company. Source: 'Creating value for Statoil and the SDFI', Statoil report presented to the minister of petroleum and energy by Statoil's board of directors, 13 August 1999. See also *Aftenposten*, 22 June 1999: 'Statoil kan bli en av verdens oiljegiganter' ('Statoil can become one of the world's oil giants').

3 The reserve replacement ratio tells us something about the relationship between exploitation of reserves and discoveries of new reserves. When the reserve replacement ratio is over 100 per cent, more new reserves have been discovered than the amount exploited.

4 *Valdez Bulletin*. Source: www.exxon.mobil.com/news/publications/valdez_bulletin/990310.html.

5 *Planet Ark*, 8 November 2001, 'US court rules $5 bln Exxon Valdez award excessive'.

6 *Planet Ark*, 8 November 2001, 'US court rules $5 bln Exxon Valdez award excessive'; *Planet Ark*, 14 January 2002, 'Exxon Valdez case to move back to Alaska court'.

7 See Exxon statement in relation to the *Exxon Valdez* 10-year anniversary. Source: www.exxon.mobil.com/news/publications/valdez bulletin; official website of the Exxon Valdez Victims: www.jomiller.com/exxonvaldez/manipulation; People of the Spill Region: www.oilspill.state.ak.us/people; Environmental Protection Agency (EPA): www.epa.gov/oilspill/exxon.

8 Personal communication with Greenpeace International, represented by Paul Horsman, Amsterdam. November 2000.

9 *Planet Ark*, 9 May 2001, 'Celebs launch UK Esso boycott over climate stance'.

10 *Planet Ark*, 12 July 2001, 'Exxon global warming boycott gets new push'.

11 *Planet Ark*, 5 July 2001, 'Esso says concerned over Body Shop's UK boycott move'.

12 Personal communication from Brian P. Flannery and Gary F. Ehlig, ExxonMobil Corporation, Irving, Texas, March 2000.

13 *Planet Ark*, 30 November 2001, 'Greens to protest at 300 Exxon UK filling stations'.

14 *Planet Ark*, 5 July 2001, 'Esso says concerned over Body Shop's UK boycott move'.

15 *Planet Ark*, 5 September 2002, 'UK poll reports switch from Esso fuel, Esso denies'.

16 Press release, 'Greenpeace to Statoil: hands off the Barents Sea', 6 March 2000. Source: www.Greenpeace.org.

17 *Valdez Bulletin*. Available online at www.exxon.mobil.com/news/publications/valdez_bulletin/990310.html.

18 Personal communication with Gerry Matthews, Shell International, Washington, DC, March 2000, and Ir. Henk J. van Wouw, Shell Nederland BV, November 2000.

19 Statoil, *Scenario Analysis* 1992/1993, cited in Estrada et al., 1997: 142.

20 Personal communication with Gerry Matthews, Shell International, Washington DC, March 2000.

21 The six divisions were exploration, international, USA, chemical, coal and minerals, and Exxon computing service.

22 Personal communication with Gerry Matthews, Shell International, Washington, DC, March 2000.

23 Greg McIvor, 'Mark of achievement: Shell's chairman lands the 1999 Tomorrow Environmental Leadership Award', *Tomorrow*, IX: 5, September/October 1999, p. 24.

24 *Financial Times*, 12 March 2002, 'A dinosaur still hunting for growth'.

25 Personal communication with Brian P. Flannery and Gary F. Ehlig, ExxonMobil Corporation, Irving, Texas, March 2000.

26 *Global Environmental Change Report*, XII: 5, p. 8: 'Lobby loses another prominent member'; personal communication with Eileen Claussen, Pew Center for Climate Change, Washington, DC, March 2000.

27 *Planet Ark*, 14 April 2000, 'Greenpeace crashes BP's party'; *Planet Ark*, 24 January 2002, 'Greens invoke profit motive for annual attack on BP'; *Financial Times*, 10 April 2000: 'Greenpeace seeks to embarrass BP Amoco over Arctic project'.

28 *Planet Ark*, 3 May 2002, 'Warming makes oil the "new tobacco"'.

29 Environment News Service (ENS) 29 May 2002, 'ExxonMobil shareholders power up renewables drive'.

30 Greg McIvor, 'Mark of achievement: Shell's chairman lands the 1999 Tomorrow Environmental Leadership Award', *Tomorrow*, IX: 5, September/October 1999, p. 24.

5

The Domestic Politics model

Company-specific differences between ExxonMobil, Shell and Statoil can shed light on differences in their climate strategies to only a limited extent. Chapter 4 revealed that company-specific features with implications for climate strategies are marked more by similarities than differences. The CA model is also incapable of explaining changes in corporate climate strategies.

We explore whether the national political contexts in which the companies operate prove more capable of explaining corporate climate strategy. As shown in chapter 2, there is reason to believe that the relationship between the companies' home-base countries and corporate strategies is important. This link will be analysed in a comparative perspective with the guidance of the Domestic Politics (DP) model. The DP model highlights the extent of social demand for environmental quality, the type of climate policy supplied by the government, and the way in which political institutions link supply and demand, that is, the relationship between state and industry. The basic assumption is that differences in corporate climate strategies can be traced back to differences along these dimensions in the home-base countries of the companies. More specifically, we assume that a high social demand, supply of an ambitious climate policy, and a link between state and industry characterised by cooperation and consensus seeking will lead to a proactive strategy. Conversely, a low social demand, a lenient climate policy, and political institutions that promote conflict and imposition are expected to lead to a reactive climate policy.

In this chapter, we focus on the Netherlands, Norway and the

US, which are the main home-base countries for Shell, Statoil and ExxonMobil, respectively. Although the oil industry is global and potentially affected by all countries in which it operates, multinational oil companies are closely tied to a specific home-base country. The significant differences observed in climate strategies between Shell, ExxonMobil and Statoil are thus possibly linked to political contexts in which these companies have their historical roots, have located their headquarters and have their main activities.

This chapter is structured as follows. In the first section, we shall explore social demand for climate policy, while the second section will take us to the actual supply of climate policy. The third section is concerned with the political institutions linking supply and demand, and the relationship between each company and its national political context is summarised in the concluding section.

Social demand for environmental and climate change protection

Social demand for environmental protection affects corporations engaged in activities associated with environmental risk. Organised social interests can influence consumer behaviour and thus be an important determinant for corporate choice by creating pressure and opportunities. Companies guard their reputations and public images mainly for economic reasons. Boycotts of Shell's petrol stations in Germany linked to the Brent Spar incident showed that losses from the boycott could cost more than the dumping alternative (Estrada et al., 1997). ExxonMobil has been concerned for its brand name as a result of the consumer boycott initiated by the green movement over Exxon's climate strategy.[1] Globalisation of communication exposes company incidents from all corners of the world. Shell's experience in Nigeria in the mid-1990s illustrates that there is nowhere to hide from negative media attention and loss of international reputation. Conversely, a high social demand for climate policy can provide market opportunities for companies by increasing 'green' consumers' willingness to pay a higher price for clean energy products. The largest Nordic energy company – the Swedish Vattenfall AB – has, for example, exploited new market opportunities by a variety of means and gained commercially from consumers' willingness to pay an extra price for clean energy (Eikeland, forthcoming). As argued in chapter 2, there is reason to assume that social demand affects corporate climate strategies *differently* depending on the companies' national ties. This is particularly true whenever national imprints overlap with market exposure.

ExxonMobil, Shell and Statoil are most directly exposed to the general public and green organisations through their petroleum product sales: in most parts of the world, consumers can choose between retail stations owned by Shell or ExxonMobil. As noted in chapter 4, the US retail market is most important to ExxonMobil. In contrast, the European market is most important to Shell, while Statoil has divided its activities between Norwegian and 'international'. In 1999, Statoil controlled 26 per cent of the market share for petrol sales in Norway, and most of its total refining capacity is located there.[2]

Against this backdrop, we assume that ExxonMobil is most exposed to a social demand for environmental protection in the US, Shell in the Netherlands and Europe, and Statoil in Norway. Based on the companies' climate strategies, we would expect that social demand – in relative terms – ranks highest in the Netherlands and Europe, lowest in the US and somewhere in between in Norway.

Several public opinion surveys have included questions about threats to the environment and attitudes to environmental protection since the 1970s. However, cross-country comparability and consistency of longitudinal data-series vary because of differences in the way questions are worded as well as in the context in which they are asked. To increase validity, we will use surveys that ask several different questions about the same phenomenon, and support public opinion surveys with interviews focusing on how social demand is actually perceived by corporate decision-makers. The organised green movement, 'green' political parties and other political parties will also be included, since they are important actors in translating public support for environmental protection into political power and climate policy. As noted in chapter 2, social demand can affect corporate strategies indirectly as well as through public policy.

The Netherlands: high social demand

With Shell's recent reorganisation, it is firmly connected with a home-base country from which corporate strategies on issues such as climate change are developed. Shell is a European company closely linked to the Netherlands. Two faces of the company appear there: Shell Netherlands BV and Shell International. Shell Netherlands possesses refineries and gas stations, and constitutes a relatively large part of the Shell Group of companies. For example, Shell's refining capacity in the Netherlands is more than twice as high as in the US. In climate policy, the main task of Shell Netherlands is to influence and implement the climate strategy of Shell International, which has the main responsibility for developing the overall climate policy for the Shell Group of companies worldwide. Shell International is located in the UK, but is heavily influenced by Dutch culture, society and policy. This influence is effectively channelled through ownership (the Shell group is 60 per cent Dutch owned), and representation on the boards of management and in the climate unit in Shell International. As pointed out by one observer, 'Shell has its backbone in the Netherlands.'[3] Moreover, national imprints overlap with market exposure in the Netherlands and Europe in terms of risks and opportunities.

A number of studies have shown that Europeans in general have been significantly more receptive to proactive policies on climate change – among both the 'elite' and the 'general public' – than have North Americans.[4] And the Netherlands is widely perceived as being among the 'greenest' countries in the world. Involvement in environmental issues, the attention paid to environmental problems, and environmentally friendly behaviour have been systematically monitored in the Netherlands since the 1980s (Bartels, 1995).[5] Up to the mid-1990s the environment was regarded as the most important social problem. From around 1994 there was a slight downward trend, when the environment shared its leading position with other issues such as crime and unemployment.

Dutch citizens are apparently willing to pay an extra price for clean energy to help combat climate change (Werf, 2000). In 1995, 60 per cent of the population in the Netherlands stated their willingness to pay higher prices for environmentally friendly products. Over 40 per cent of the population was willing to pay higher taxes for better environmental quality and was even willing to accept a lower standard of living.[6] Involvement of citizens in the climate-change issue has varied according to publicity for specific events (Schenkel, 1998).[7] A peak occurred in 1992 related to the Rio conference (almost 40 per cent involved). The corresponding figure in 1996 was slightly above 30 per cent (VROM, 1997).

The relatively high social demand has led Shell to use the Netherlands as a test ground for society's willingness to pay for environmental protection in general, and clean energy in particular.[8] In essence, the Netherlands serves as a sort of playground for Shell's investment in renewable technology. Inglehart (1971) has proposed that the environmental movement represents a political expression of post-materialistic values that will be strengthened as the new generations become older. Weale (1992: 138) claims that a high proportion of the population in the Netherlands is influenced by post-materialistic values, in Inglehart's terminology. This underlying trend contributes to making the Netherlands an interesting test case for future demands for clean energy.

Time-series data on membership and income patterns among the most important Dutch ENGOs show a steady increase up to the mid-1990s (Skjærseth, 1999). This peak corresponds with Shell's process of reorganisation initiated in the mid-1990s. The Dutch green parties have had a relatively stable and high electoral basis since 1984. In 1989, the Green Left won six parliamentary seats out of 150 and doubled the number of seats of the constituent parties. The 1994 election left them with five seats. In 1998, political parties with green ties increased their share of seats in parliament, thus indicating that the decline in Dutch awareness has been modest. For example, the Green Left more than doubled its share of seats in parliament to 11 (Luciarde, 1999). The sensitivity of other political parties partly explains why the Dutch greens have not had even stronger support. Public perception of environmental issues has little to do with voting behaviour in the Netherlands (Tak, 1994: 11). An analysis of the parties' programmes shows that all major political parties have adopted green ideas, and do not differ significantly in their rhetorical support for environmental protection.

The development in Dutch social demand for environmental and climate policy appears in line with expectations. The climate strategy of Shell accords well with social demand in the Netherlands. Moreover, Europeans have generally been more receptive than the US to an active climate policy. The Dutch home-country context of Shell constitutes a population where a large share of the consumers have signalled their willingness to pay higher prices for environment-friendly products and clean energy. This has led Shell to use the Netherlands as a test country for predicting future social demand. Conversely, Shell's petroleum product sales in Europe make the company vulnerable to negative attention and consumer campaigns.

The United States: low social demand

ExxonMobil is deeply rooted in the US even though the company operates all over the world. Exxon Corporation started out as Standard Oil in 1882, and the corporation's headquarters is located in Irving, Texas, where corporate staff develops Exxon's environmental strategy. Exxon's US cultural heritage has been seen as important for the company's choice of environmental strategy (Estrada et al., 1997). Like the case for Shell, national imprints overlap with ExxonMobil's market exposure in the US.

North Americans express significant concern for the environment, but climate change is given little public attention compared to other environmental problems.

Table 5.1 *Percentage expressing 'a great deal' of concern in the US*

Issue	1989	1999	2000	Change since 1989
Pollution of drinking water	–	68	72	–
Pollution of rivers, lakes and reservoirs	72	61	66	–6
Contamination by toxic wastes	69	63	64	–5
Ocean and beach pollution	60	50	54	–6
Air pollution	63	52	59	–4
Contamination by radioactivity	54	48	52	–2
Loss of habitat for wildlife	58	51	51	–7
Damage to the ozone layer	51	44	49	–2
Loss of tropical rain forest	42	49	51	+9
Global warming	35	34	40	+5
Acid rain	41	29	34	–7

Source: Based on Saad and Dunlap (2000).

Data displayed in table 5.1 are not directly comparable with Dutch data. This means that the percentage expressing 'a great deal' of concern in the US cannot be directly compared to the Dutch measure of 'involvement'. However, table 5.1 provides information on trends and the rating of climate change as an issue within the US. First, of the eleven issues included here for which longitudinal data exist, public concern has been reduced for eight. We should bear in mind, however, that public concern for the environment ranked particularly high in 1989 owing to, among other things, a high level of media attention linked to the March 1989 *Exxon Valdez* oil spill – the largest in US history (see chapter 4). This Exxon-caused incident contributed to raising the social demand for environmental protection in the US (Saad and Dunlap, 2000). Second, there have been significant fluctuations the last decade. This indicates a pattern in line with 'issue-attention' cycles (Downs, 1972), which provide companies with an ambiguous social context. Thirdly, global warming is located at the bottom of the spectrum in spite of an increased concern between 1999 and 2000. A study of media coverage of climate change follows roughly the same pattern and shows significant fluctuations with a peak in 1997, when the Kyoto Protocol was signed (Agrawala and Andresen, 1999).[9]

The US public worries about national environmental issues such as water and air pollution as well as toxic waste, rather than international ones such as climate change (OECD, 1996). A Gallup poll released in connection with the Kyoto negotiations indicates that North Americans are somewhat inconsistent in their attitudes to climate change (Gallup and Saad, 1997). On the one hand, a majority believes that the problem will have harmful effects within the next 25 years. On the other hand, they are not willing to accept significant costs or a large share of the international burden to reduce the problem. This indicates a different attitude among US consumers and those in the Netherlands. While Shell perceives the Dutch consumers' willingness to pay for clean energy as a business opportunity, fossil-fuel interests in the US have launched PR campaigns to highlight the high economic costs for consumers resulting from GHG emissions reductions. For example, Exxon Education Foundation's *Exxon Energy Cube*, with videos, books, games and posters, 'implies that fossil fuels in general pose few environmental problems and that alternative energy is unattainable and costly' (Levy and Egan, 1998: 350).

According to Skolnikoff (1997), the role of public opinion in the US is unclear because climate change is not a major issue on the public agenda. Social 'demand' for a stringent climate policy generated from 'below' appears to be a misleading description of the US situation. The Clinton administration did in fact make an effort to create such a demand through various education campaigns. A number of governmental initiatives under the heading 'public outreach' were launched in the 1990s aimed at increasing public awareness by a number of means, from multi-media presentations to displays in shopping malls (DSP, 1997).

With a low public demand for climate policy and a low willingness to pay for clean energy, ExxonMobil has not had strong incentives to exploit market opportunities related to renewables in the US. Instead, the company has sought to keep demand for stricter climate policy low through PR campaigns. According to the company, oil and gas are significantly more profitable than renewables.[10] ExxonMobil also maintains that the company has poor experience with renewables: in the 1970s, Exxon, like many other companies such as Shell, invested unsuccessfully in renewables like solar energy.

US ENGOs reflect the difference in attention between domestic and international environmental problems. On the one hand, green groups are a vital force in American society with respect to domestic problems: 16 per cent of Americans say they are active in the environmental movement, while 5 per cent indicate a membership in large national and international organisations such as the Sierra Club and Greenpeace. Compared to other social groups, 43 per cent strongly agree with the goals of the environmental movement, which places it as number three of eight major social movements after the civil and women's rights movements. In terms of perceived impact on policy-making, however, green groups clearly lag behind two additional movements: gun control and abortion rights (Dunlap, 2000).

On the other hand, while climate change has top priority in Greenpeace International based in Amsterdam, Greenpeace-US in Washington, DC, finds it difficult to raise funding for campaigns for the Kyoto Protocol, owing to low public concern. Even though US ratification of the Kyoto Protocol, would have been the key to a viable climate regime, Greenpeace-US resigned from its efforts to lobby for US ratification even *before* George W. Bush was elected president in the autumn of 2000.[11] It is quite illustrative that the Stop Esso campaign specifically targeting ExxonMobil's climate strategy was initiated in Europe and not in the US (see chapter 4). In July 2002, a Greenpeace-commissioned survey, undertaken by the polling agency MORI, showed that a significant number of British motorists have stopped buying petrol from Esso stations and have switched to other retailers as a result of the Stop Esso campaign. Esso, however, denies the result and says that the company's business has not been affected.[12] In essence, the green groups are able to exploit a higher level of political concern in Europe that may affect the market shares of ExxonMobil.

The US party system is significantly different from those of most European states, including Norway and the Netherlands, owing to the dominance of two parties and the lack of proportional representation. Crudely put, the environment is typically a liberal concern in the US, and thus advocated by the Democrats. Republicans frequently oppose environmental legislation since an active environmental policy is perceived to hamper economic growth, raise unemployment and cause more government. The two-party 'winner-takes-all' system prevents environmental issues from becoming major political issues in the US. Environmental issues have seldom been a major factor in national elections in the US: environmental protection was ranked as issue number eight in a recent Gallup poll (Saad and Dunlap, 2000). Norway and the Netherlands have both experienced the resignation of governments triggered by a climate-related issue – events unlikely to take place in the US (see below).

Social demand for climate policy in the US appears somewhat ambiguous, but corresponds roughly with expectations. On the one hand, a large proportion of the US public expresses concern for the environment, and ENGOs are a vital force in US society. On the other hand, the US public worries mainly about national environmental issues and does not accept significant costs to deal with climate change. In addition, environmental protection, including climate change, ranks low on the political agenda. The social pressure exerted on ExxonMobil concerning climate change thus appears relatively low. International ENGOs such as

Greenpeace gear their resources towards the European face of ExxonMobil: Esso. Equally important is the North Americans' unwillingness to pay a higher price for clean energy. The fossil-fuel lobby has exploited these attitudes, and ExxonMobil does not perceive investments in renewable energy sources as an interesting business opportunity.

Norway: fluctuating demand

Statoil has concentrated most of its operations on the Norwegian continental shelf, and it plays a crucial role in the Norwegian economy: in 1998, the Norwegian petroleum sector accounted for 11.8 per cent of GDP and 29.8 per cent of total export (Andersen and Austvik, 2000). Statoil is a significant supplier of natural gas to Europe and the largest retailer of petroleum products in Scandinavia. Statoil's headquarters are located in Stavanger, Norway. The link between Norway and Statoil is thus very strong and direct: 'Statoil's own trademark is to a large extent Norway's trademark' (Estrada et al., 1997: 149). The overlap between national imprints and market exposure is thus extremely strong in the case of Statoil.

Norwegian attitudes to climate policy and the environment have fluctuated significantly over time. Surveys on the importance of environment and energy issues conducted as part of national election research show quite dramatic variation.[13] In 1989, the environment and energy ranked as the second most important political issue overall, and 37 per cent considered it the most important. By 1993, the environment and energy issue had dropped to number five, and it was considered most important by only 7 per cent (Aardal and Valen, 1995). The significant increase in interest in the latter part of the 1980s and the decline in 1993 are also confirmed by a number of other surveys (Skjærseth, 1999). Furthermore, the level of public concern witnessed from 1993 seems to have stabilised at a low level. For example, in 1990, 23.5 per cent thought that public expenditures on the environment should be increased significantly, while the corresponding figure in 1996 was 8.8 per cent (Skjåk and Bøyum, 1996).

Less concern for environmental problems has also been revealed by a study called Norsk Monitor.[14] In 1989, 61 per cent characterised the situation as 'grave' and agreed with the need for 'drastic action'. The corresponding figures in 1997 and 1999 were

34 per cent and 28 per cent respectively. Of five environmental problems, global warming is ranked after ozone depletion and acid rain in the same study. However, the population's fear of climate change has decreased significantly: in 1989, 40 per cent of the population was very worried, while only 22 per cent was worried in 1997. In short, Norwegian developments seem largely in line with the patterns following from Downs' (1972) idea of issue-attention cycles. Statoil has been exposed to more ambiguous demand than Shell in the Netherlands. Moreover, Norway relies heavily on renewable hydroelectric power at the outset and electricity prices have been low. This makes the market for solar, wind, wave or biomass very small and efforts to introduce 'green' electricity almost futile. Statoil does not see any great market opportunities for these energy carriers. In January 2000, Statoil withdrew from a joint venture on wind power, arguing that it could not see any future profitability for wind power in Norway.[15]

The largest ENGO in Norway, Naturvernforbundet (Society for the Conservation of Nature), decreased significantly in both membership and income between 1993 and 1995. Naturvernforbund has since remained small in comparison to the beginning of the 1990s. The growth and decline of the organisation illustrate a general trend in membership among the most important Norwegian ENGOs (Jansen and Osland, 1996). Statoil has generally been influenced by Norwegian ENGOs, but the company has not been exposed to any serious threats of boycotts over its climate policy from the green movement in Norway. However, ExxonMobil has been targeted in Norway as part of the Stop Esso campaign.[16]

There is no truly 'green' party in Norway. This phenomenon has been explained by the general sensitivity of the political system to new social demands. Existing political parties have largely managed to absorb a potential 'green' party electorate. As in the Netherlands, few voters perceive environmental issues as important for party choice, except for two small parties – the Socialist Left and the Liberal Party. These have to some extent filled the niche for a 'green' party (Seippel and Lafferty, 1996).

Public opinion data in Norway are not directly comparable with either Dutch or US data. However, we have a strong impression that Norwegian demands have been the least stable during the 1990s. There has been some social pressure mediated by ENGOs on Statoil's climate policy, but market opportunities for new renewable energy have been perceived as minor. Solely on the basis of social demand, we would expect the climate strategy of Statoil to be more in line with ExxonMobil's than Shell's.

Comparison of social demand

Even though comparison of social demand across countries is difficult, because different questions are asked in different contexts, we can draw some conclusions about relative trends and differences. First, a dividing line seems to go between the US and Europe: public opinion, green organisations and political parties exert a stronger pressure for climate change measures in Europe than they do in the US. The most visible expression of this phenomenon is perhaps the resignation of governments over climate-related issues in both the Netherlands and Norway – incidents that are unlikely to occur in the US. Second, there are significant differences within Europe, as reflected in those between Norway and the Netherlands. Social demand for climate policy appears to be higher and more stable in the Netherlands than in Norway. This means that all three cases roughly support our expectation of a close link between social demand in their home-base countries and corporate climate strategies: the home-base country context of public pressure and opportunities varies in accordance with the climate strategies of Shell, ExxonMobil and Statoil.

However, causal relationships between social demand and corporate climate strategy are more difficult to establish than the correlations witnessed above. Shell uses the Netherlands as a test country for the future of opportunities for clean energy. Thus, social demand in the Netherlands and Europe seems to be one important explanatory factor for Shell's proactive climate strategy. In the case of Statoil, the extreme importance of Norway and Norwegian markets has apparently led the company to follow fluctuations in public opinion. Lack of viable market opportunities for renewable energy other than hydroelectric power, combined with only moderate social pressure, can shed some light on Statoil's climate strategy. ExxonMobil has been described as a super-tanker: steady and strong. Currently, pressures and opportunities signalled from US society do not provide ExxonMobil with sufficient incentives to change course. However, the European market is also important to ExxonMobil, and the company has recently been exposed to substantial campaigns in Europe explicitly linked to its climate strategy. ExxonMobil appears less vulnerable to European pressures than Shell – at least for the time being.

All in all, we can conclude that social demand in the companies' home-base countries corresponds with expectations derived from the DP model. However, social demand represents only a part of the total causal picture. In the next section, we will look at the supply side, i.e. the link between governmental climate policy and corporate climate strategies. To various degrees, climate policy is likely to reflect social demand in democratic societies. Social demand can thus also affect corporate strategy *indirectly* through public policy. However, this link also depends on the respective political institutions' sensitivity to new social demands. While environmental concerns have been quickly absorbed into the multi-party political systems of Norway and the Netherlands, these issues apparently have more difficulties in penetrating the US two-party system.

Governmental supply of climate policy

This section aims to explain differences in corporate strategies from the perspective of governmental supply of climate policy. The main focus is on how national climate policy affects the oil industry rather than why the policy itself changes and varies.[17] Corporate response is likely to depend on the level of ambition of the climate policy measured in terms of targets and policy instruments. A viable climate policy creates regulatory pressure, grants market opportunities and reduces uncertainty for companies with regard to future governmental priorities.

Ambitious GHG reduction targets linked to mandatory policy instruments send a clear signal to industry. In such situations, governments show that the problem is taken seriously and action is expected at the level of target groups. Company response can thus reflect a desire to avoid further costs of governmental regulation. If these targets are combined with an ambitious governmental policy on renewable energy, stimulating market opportunities, companies can be expected to respond proactively.

Conversely, a situation characterised by lenient GHG targets, voluntary public programmes and low priority for renewables is likely to spur a reactive response whenever the climate issue is perceived as a potential threat to business interests. A combination of high social demand and an ambitious climate policy will create a positive interplay between factors pulling in the direction of proactive climate strategies.

Against the backdrop of the corporate climate strategies observed in chapter 3, we assume that the Netherlands has adopted the most ambitious climate policy, followed by Norway and the US – in that order. The term 'ambitious' should be understood in relative terms only, not according to the actual requirements of the problem at hand.

In the following section, we have distinguished between overall climate policy and targets, and policy instruments directly affecting the oil industry.

The US: low ambition

Overall climate policy   The US is the single largest contributor to global GHG emissions (about 25 per cent of the global total). Accordingly, domestic US efforts to combat climate change would make a significant difference in terms of global anthropogenic emissions. A Kyoto agreement without the United States was until the 2001 Bonn Summit widely perceived merely as a theoretical possibility.[18] This means that US climate policy during the 1990s sent an extremely important signal about the road ahead to the US oil industry and ExxonMobil.

US climate policy can be divided into three phases, corresponding roughly with changes in administrations and international developments. The first phase covers the period up to the 1992 Rio Summit and the last year of the Bush–Quayle administration. The second commences with the Clinton–Gore administration in 1993 and ends before the run-up to the Kyoto conference in 1997. The third runs from the Kyoto negotiations to the Bush–Cheney administration.

Spurred by the drought and heat waves that hit the US in 1987 and 1988, the Bush–Quayle administration initially expressed deep concern about threats to the global climate. However, initial enthusiasm rapidly declined, and the administration became critical of the findings of the IPCC's first assessment report (Bergesen et al., 1995). As this was combined with fear of large socio-economic costs that could threaten 'the American way of life', the US was reluctant to support an international agreement including 'targets and timetables' during the negotiations for a climate convention in the Intergovernmental Negotiating Committee (INC) that commenced in 1991.[19] The US preferred a comprehensive and flexible approach – one that would include the sources and sinks of all GHGs within the framework of any climate agreement (Agrawala and Andresen, 1999).[20] This position is usually referred to as the 'bottom-up' approach, emphasising action from below, in contrast to the 'top-down' targets-and-timetables approach preferred by most European states.[21] Having succeeded in keeping targets and timetables out of the UNFCCC, the US, as one of the first countries, both signed and ratified the convention in 1992. This indicates a close match between national interests and the final international output.

The Clinton–Gore administration took office in January 1993 and initiated the second phase of US climate policy. This is characterised, first, by the adoption of a unilateral target. In April 1993, Clinton announced that the US had committed itself to reducing emissions of greenhouse gases to their 1990 levels by the year 2000. The Clinton–Gore administration also presented an action plan on how to reach the stabilisation target. A British thermal unit (BTU) tax was initially proposed based on the heat content of the fuel. This tax aimed to stimulate energy efficiency and cut federal deficit by raising about US$72 billion in tax revenues over five years. The tax was, however, turned down by the Senate in spite of a Democratic majority in Congress. The energy tax provoked the US oil industry, which played a crucial role in killing it.

Shortly after the tax defeat, the Clinton administration announced the 1993 Climate Change Action Plan (CCAP), which aimed to increase energy efficiency in various sectors. In 1997, the US followed up the CCAP with the United States Climate Action Report (USCAR) to the UNFCCC. These plans aimed at tapping the large potential for reducing GHGs in the US by means of 'no regrets' measures.

CCAP and USCAR consist of more than 40 actions that are to be implemented primarily by public voluntary programmes, information campaigns and partnerships between business and government. More than 5,000 organisations from around the country participate in CCAP/USCAR programmes. Most of these programmes represent 'no regrets' measures that focus on *technological* innovation and use. The programmes seek to create markets for investments in existing, or close to existing, technologies capable of reducing emissions. Thus, the principal task of the government was to stimulate innovation through voluntary programmes and correct market failures by means of information and persuasion.

There has been a heated debate between different departments and agencies on the effectiveness of the climate change programmes (OECD, 1999). However, the CCAP did not achieve, or even come close to achieving, the stabilisation target. Between 1990 and 2000, US GHG emissions increased by 14.5 per cent (EPA, 2002). In fact, the stabilisation target was never taken seriously by the US oil industry. First, the plans significantly underestimated the reductions needed to return emissions to 1990 levels by the year 2000. Second, the plans were not fully funded. Republicans had a majority in Congress from 1995 and generally resisted new taxes or increased spending.[22] Third, the plans had very low political priority. The priority of the US during the 1990s was research, not action (Brunner and Klein, 1998).

The Clinton–Gore administration supported the second IPCC assessment report. Since the late 1950s, US federally supported science has been the single most important cause in identifying climate change as a global problem. The US led the establishment of the IPCC, which provided the scientific background in the form of assessment reports upon which both the UNFCCC and the Kyoto Protocol rest (see chapter 6). However, US scientists remain split on this issue and some even deny the validity of the IPCC analysis altogether. This is particularly important for US positions and policy, since the US tradition allows more scientific involvement in policy matters than is the case in Europe (Skolnikoff, 1997).

After 1995, the Republicans controlled both the Senate and the House of Representatives in Congress. Nevertheless, the US government agreed to, and signed, the 1997 Kyoto Protocol, which calls upon the US to reduce GHG emissions by 7 per cent from 1990 levels. While the US fossil-fuel industry, with Exxon as one of the most prominent leaders, to a large extent controlled the development of national US climate policy, this move by the Clinton–Gore administration indicates that the international process developed beyond the control of ExxonMobil. The US Kyoto commitment implied a dramatic strengthening of US climate policy. Analysis showed that the US would have to reduce emissions by the order of 30 per cent relative to a 'business as usual' scenario to reach the Kyoto target.[23] However, the Senate would have to ratify the Kyoto Protocol before it could be viewed as a part of US policy (see chapter 6).

Four years later, in January 2001, the former Texas governor George W. Bush was elected president. In March 2001, Bush Jr declared that the Kyoto Protocol was unacceptable because it would harm the US economy and because it failed to hold developing countries to strict emission limits. In February 2002, Bush unveiled proposals for a voluntary scheme to curb GHG emissions.[24] This represented a continuation of the voluntary approach under the Clinton–Gore administration, bringing US climate policy back to square one. In retrospect, US climate policy may over time be characterised as relatively weak. This pattern was, however, severely disrupted by the US Kyoto commitments lasting from 1997 to 2001.

Policy instruments and the oil industry   The 1993 CCAP and 1997 USCAR do not include programmes directly targeting the oil industry, but comprise general cross-sector programmes that affect this industry together with others. 'Green Lights and Energy Star', 'Climate Wise' and 'Climate Challenge' are among the most important programmes.[25] For example, 'Climate Wise' focuses on the industrial sector, which accounts for about 30 per cent of US energy consumption. The programme helps companies realise their energy efficiency potential by providing technical assistance and public recognition. Each participating company is to develop an action plan within six months and report results of its actions annually.

Company members of the API participate in various voluntary programmes including 'Climate Wise', 'Green Lights' and 'Natural Gas Star'. ExxonMobil has generally shown little interest in the programmes, arguing that none of them has led the company to take action that departs from what it would have

done in their absence.[26] However, ExxonMobil and Shell participate in the 'Natural Gas Star' programme. According to the
Environmental Protection Agency (EPA), this programme has
been an outright success. By the year 2010, it aims at energy cost
savings in the order of US$100 million and methane savings of 55
billion cubic feet.[27] As of spring 1999, the programme included
17 production partners and 53 transmission and distribution
partners. The partners have exceeded CCAP goals by preventing
the release of 75.8 billion cubic feet of methane gas, valued at
about US$152 million. The producers have accounted for 51
billion cubic feet.[28]

The US oil industry is also heavily subsidised. For national,
economic and energy security reasons, the government is strongly
involved in this sector. The defence of Persian Gulf oil supplies
and the commitment to maintain a government-owned strategic
petroleum reserve constitute two important (energy) policy goals.
The introduction of additional supplies into the market is seen as
an effective way to dampen the price rise and mitigate the
economic damage resulting from severe oil supply disruption.
These and other subsidies are provided to producers, transporters
and consumers. According to Greenpeace, subsidies in domestic
oil are worth between US$1.20 and US$2.80 per barrel of domestic crude consumed. Cutting subsidies would thus represent an
effective climate-change strategy (Koplow and Martin, 1999).

In sum, ExxonMobil was during the 1990s exposed to weak
and ambiguous climate-policy targets in the US. ExxonMobil did
not take the US stabilisation target seriously, but the US Kyoto
commitments represented a temporary change in ambitions. At
the policy instrument level, ExxonMobil has voluntarily participated in public programmes, though without much enthusiasm.
In short, ExxonMobil has been exposed to a national climate
policy context characterised by high political uncertainty and
little regulatory pressure.

The Netherlands: High ambition

Overall climate policy   Because of a concern for sea-level rise,
the Netherlands was perhaps the only OECD country that
responded ambitiously to climate change throughout the 1990s
(VROM, 1999).[29] For Shell, the Netherlands not only constitutes

a 'test country' for pressures and opportunities offered by the general public but also represents a test case for what the industry can expect from a relatively viable climate policy.

In November 1989, the Dutch government announced its decision to stabilise CO_2 emissions at the 1989/1990 level by 1995 at the latest. In 1990, a revised plan called for a 3–5 per cent reduction from average 1989/1990 levels by 2000. In 1995, the CO_2 target was reformulated to a 3 per cent reduction from the 1990 level by 2000. In 2002, the Netherlands was the first of the EU member states to ratify the Kyoto Protocol. Simultaneously, policy instruments have been gradually stepped up. This has primarily taken place in the various National Environmental Policy Plans (NEPPs), as well as in a number of white papers covering other sectors of society such as energy, transport, agriculture and waste (VROM, 1989, 1993, 1998). The Netherlands is renowned for its comprehensive environmental planning, and climate policy is no exemption: between 1993 and 1997 the Dutch authorities published about 30 policy documents with relevance for climate change (VROM, 1997). In 1999, three new packages of policy instruments and measures were proposed (VROM, 1999). In spite of these efforts, CO_2 emissions grew by about 11 per cent between 1990 and 1997 due to lower energy prices and higher economic growth than expected.

There are no distinctive phases in Dutch climate policy. At the domestic level, the Netherlands has gradually stepped up its efforts over time. The Dutch position has traditionally been that domestic reductions are most important, and the main reduction of GHGs should come from the industrialised countries (IEA, 1994). The Netherlands has therefore been in a good position to be in the forefront at the EU and the global levels.

Climate change was first mentioned domestically at the beginning of the 1980s. The need to offset the expected sea-level rise was placed on the agenda by a report published by the Public Health Council in 1986 – the year that experts started to prepare the report *Concern for Tomorrow*.[30] This report was released in 1988 and alerted the general public and politicians alike by painting a dark picture of the state of the Dutch environment. Because of its scientific credibility, the report had a profound impact (Bennett, 1991). Moreover, the rising tide of public opinion affected the political climate (Weale, 1992). In the 1989 election, the environment in general and climate change in particular became prime issues. The first NEPP was released in 1989. NEPP 1 shifted the focus from regulations and standards to negotiated agreements between the government and sector target groups. In the field of climate change, the target was set to stabilisation of CO_2 emissions by 1995 at their 1989/1990 level. The main policy instrument to achieve this goal was negotiated agreements with specific industrial target groups and other sectors.

NEPP 1 is probably best known for triggering the fall of the first Dutch cabinet on the grounds of an environmental issue. The Christian Democratic/Liberal coalition had to resign in 1989 because the Liberals resisted raising the costs of car driving. The new government – a coalition of Christian Democrats and the Labour Party – published a new version of NEPP in 1990: NEPP Plus. NEPP Plus went further than NEPP 1 by adopting the 3–5 per cent reduction target to be achieved by 2000 (VROM, 1991).

NEPP 2 was released in 1993 and signalled that the objectives of NEPP Plus would continue to apply (VROM, 1993). NEPP 2 placed more emphasis on strengthening implementation, particularly with regard to 'diffuse' sources, and target groups that were difficult to reach by means of negotiated agreements. An energy tax was proposed to reach such groups. The government preferred to work vigorously for an EU-wide tax, but was prepared to adopt a national tax if the EU tax did not materialise. A regulatory energy tax for small consumers was adopted in 1996. The green tax is regulatory since it aims explicitly at reducing consumption. It applies to natural gas and electricity consumption and comes in addition to the environmental tax on all fossil fuels.[31] The tax raises energy prices for small-scale consumption by 15–20 per cent. Renewable energy is exempt from the tax. The small-consumer tax is expect to contribute to a total CO_2 emissions reduction of the order of 1.5–5 per cent (Baron, 1996).

In the 1995 third White Paper on energy, the aim was to increase the share of renewable energy to 10 per cent of total energy consumption and to improve energy efficiency by one-third in the year 2020 compared to 1990. The current share of renewables in the Netherlands was about 1.5 per cent. Over the preceding few years, policies to promote renewables had been strengthened by the introduction of a wide range of fiscal arrangements, such as energy investment tax credits. Renewable energy was also promoted through its inclusion in the long-term agreements (VROM, 1999). The 1995 and 1996 goals on renewables directly influenced Shell's decision to establish Shell International Renewables in 1997.[32]

The third NEPP issued in 1998 followed the same path as NEPP 2 (VROM, 1998). The main message was to strengthen policy instruments and measures in order to reach the targets adopted in NEPP Plus. With respect to climate change, the Cabinet intended to increase energy taxes except for heavy energy consumers. In 1999, the Ministry of Housing, Spatial Planning and the Environment (VROM) presented a new climate policy implementation plan (VROM, 1999). The plan was developed in response to scenarios showing that GHG emissions, particularly of CO_2, would continue to grow under current policies. The new plan was based on a package approach. The Netherlands assumed that about half of the cutbacks could be dealt with through the flexibility mechanisms in the Kyoto Protocol. The *basic package* consisted of measures and instruments expected to bring about the necessary reductions, excluding the share under the flexibility mechanisms. Increasing the share of renewable energy, further improvement of energy efficiency, and measures taken at coal-fired power plants were the main strategies for the future. The *reserve package* was designed to reduce emissions quickly if the basic package failed. The core of this package consisted of raising the regulatory energy tax and excise duties on motor fuels, reducing N_2O from emissions in the chemical industry and storing CO_2 underground. Implementation of this package required future political decisions. In addition, an *innovation package* was prepared primarily for continuous reduction beyond the Kyoto horizon. This package aimed to promote innovation in technology and in governmental policy instruments that might stimulate GHG reductions.

The EU member states and the Commission have tried to hammer out some sort of burden sharing within the EU since the EU adopted its stabilisation target in 1990. Particularly, Greece, Ireland, Portugal and Spain feared that EU climate policy could hamper their economic development. Earlier efforts were, however, unsuccessful until the Netherlands assumed the

EU half-year rotating presidency in January 1997. The internal EU negotiations were part of the global negotiation process – in which the EU had strong leadership ambitions. The Netherlands played an important role in this process by proposing the 'Triptique Approach' (Ringius, 1997).[33] In contrast to the previous targets- and-measures view, this new approach paved the way for a binding EU agreement finalised in June 1998. The EU proposed, first, that the OECD countries should cut their GHG emissions by 15 per cent in 2010. Second, the Council set forth a burden- sharing agreement including emissions targets for each member state. While some states were allowed to increase their emissions, the Netherlands was required to reduce its by 10 per cent in 2010 relative to 1990 levels.[34] In short, the Netherlands sent a clear message to industry that contains both pressures and opportunities.

Policy instruments and the oil industry   The Dutch oil industry is beginning to accept the reality of climate change and to consider measures in response. Dutch and EU initiatives on renewable energy have had a significant impact on Shell's climate strategy. The company believes that renewables will become the main energy source in the future and perceives them as an interesting business opportunity (see chapter 3). Accordingly, Shell seeks to create a new image as an energy company with increasing activities in solar power and biomass.

However, the main policy instrument applying specifically to the oil and gas production sector is a long-term agreement (LTA) that was concluded between the authorities and 12 companies and ventures in 1996. One of these is Nederlandse Aardolie Maatschappij BV, which is a 50–50 joint venture between Dutch Shell and Exxon. This agreement sets the target for energy efficiency improvement over the period 1989-2000 to 20 per cent. The measures needed to achieve this objective are set out in a long-term plan for improvement of energy efficiency. This is confidential, but all such plans include the following elements (Nuijen, 1999):

- an assessment of energy consumption in the reference year 1989;
- a survey of opportunities for energy efficiency improvement;

- a model for company energy plans;
- monitoring and energy management in each company;
- research and development on new low-energy technologies;
- demonstration projects for energy-saving measures;
- market introduction of low-energy techniques;
- assistance to individual companies;
- transfer of know-how and information.

The oil and gas production sector has improved energy efficiency in the field of transport, oil pumps and processes. Nearly half the improvements stem from process modification in order to recover cooling water (Novem, 1999).

Negotiation of an LTA typically takes from one to two years, from a letter of intent to signature. The impact of the LTA is very hard to assess, but table 5.2 indicates that a large part of the improvement would have taken place in any case: before 1996, 18 per cent had already been achieved in the absence of the LTA. On the other hand, the largest share of improvement in energy efficiency occurred after the signing of the agreement.

Table 5.2   *Improvement of energy efficiency in Dutch oil and gas production*

Year	Energy efficiency index [a]	Energy efficiency improvement in TJ/year [b]
1989	100	0
1990	99	347
1991	98	359
1992	94	947
1993	92	1181
1994	86	2419
1995	82	1749
1996	80	1178
1997	75	3710
1998	70	3849

Notes: [a] The definition of the energy efficiency index is the energy consumption in the year in question to produce the total output in that year, divided by the energy consumption that would have resulted had the same production been made with energy efficiency in the year of reference (1989).
[b] In absolute terms.
Source: Based on Novem (1998) and Nogepa (1998).

Oil companies are also affected by general measures to increase energy efficiency by, for example, increasing the use of co-generation. Natural gas is the dominant primary energy source in the Netherlands, with a share of about 50 per cent in total primary energy consumption, higher than any other International Energy Agency (IEA) country. Oil accounted for 36 per cent, coal for 12 per cent and nuclear energy for 2 per cent, while 1 per cent was net electricity imports (Slingerland, 1997). Due to the large share of natural gas, decentralised CHP has experienced a major boom since the late 1980s, partly due to climate policy: from a stable 1,500 MW for many years to over 4,500 in 1995. The largest part consists of industrial CHP. There has also been a shift to gas in the fuel mix.

Gasuine is the venture responsible for the distribution of natural gas to electricity generation companies, distributors and some large industrial end users. It is partly state owned and partly owned by Exxon and Shell (Slingerland, 1997). In 1990, the distributors signed an agreement with the authorities that included an environmental action plan. This plan specified measures to be taken by the distribution companies that aimed to reduce CO_2 emissions by 9 million tonnes by the year 2000. In 1994, as a result of the second memorandum on energy conservation and NEPP 2, a new target was set at 17 million tonnes of CO_2.

The next generation of LTAs spans the period 2000–2010. Since the most obvious measures were taken from 1989 to 2000, the range of measures and themes has been extended by focusing more closely on product efficiency and industrial cooperation across themes, such as transport. For energy-intensive industry – including refineries – the LTAs will be replaced by so-called benchmark agreements (BA). The underlying logic is that Dutch energy-intensive industry cannot be pushed further than to become and remain the best in the world in terms of energy efficiency. If companies do not comply, the 1996 regulatory tax may be expanded to include energy-intensive industry.

Summing up, Dutch climate policy in the 1990s was characterised by relatively ambitious targets and increasingly vigorous policy instruments. Over time, this policy reduced uncertainty and affected Shell's climate strategy in two ways. First, Shell has been pressured by regulation on energy efficiency and negotiated agreements backed up by threats of regulatory taxes. Second, and more importantly, the company has been 'offered' new market opportunities by Dutch and EU policy on renewables. Notice that Dutch climate policy did not affect ExxonMobil's climate strategy even though ExxonMobil in the Netherlands has been exposed to the same climate policy as Shell. On the other hand, ExxonMobil has adapted to the climate policy in the Netherlands.

Norway: ambiguous ambition

Overall climate policy   In the late 1980s, a green wave of public concern for the environment washed over Norway. In 1987, the UN Commission on Sustainable Development – led by the then prime minister, Gro Harlem Brundtland – released the Brundtland Report. The report, entitled *Our Common Future*, emphasised climate change as a major problem. Thus it came as no surprise that Norway was the first country in the world to adopt a unilateral stabilisation target for CO_2 at 1989 levels by 2000. Moreover, it was one of the first countries to introduce a CO_2 tax covering mainly off-shore activities in 1991. This was met with strong opposition from the entire oil industry – including Statoil – even though Statoil was a fully state-owned company at the time.

Initial political enthusiasm, however, was soon replaced by pragmatic economic concerns. As a petroleum exporter, Norway was expecting a steep increase in its emissions of CO_2 from petroleum activities (of the order of 60 per cent from 1989 to 2000), and stringent policy instruments could affect petroleum markets and export: almost one-third of total Norwegian export income came from petroleum export (Sydnes, 1996). Moreover, Norwegian energy production is based largely on hydroelectric power, which limits the country's potential to reduce emissions by changing energy consumption patterns and leaves the petroleum sector as one of the largest sources of domestic CO_2 emissions. The structure of Norway's energy consumption and its dependence on petroleum exports led the country to seek solutions abroad rather than at home.

Norwegian climate policy can be divided into three phases. In the first phase, from the 1989 target to the 1992 Rio Summit, Norway developed a predominantly domestic strategy to combat climate change, linked with a high international profile. In the second phase, from 1992 to the Kyoto negotiations, the country's ambitions to combat GHG emissions domestically were significantly reduced. In the last phase, from Kyoto and beyond, Norway first adopted a more balanced strategy, acknowledging the need for domestic cuts in emissions combined with reductions abroad through flexible international arrangements. Thereafter, this policy drifted away from domestic cutbacks, before a national quota system that aimed to reduce domestic emissions was proposed in 2002.

The debate on a Norwegian CO_2 target started in 1987 and culminated in the stabilisation target adopted in 1989. The ambitious target came as a result of a 'green beauty contest' between the political parties in the Storting (the Norwegian parliament) (Bergesen et al., 1995).[35] With the adoption of the CO_2 tax, Norway kept a high international profile during the INC negotiations starting in early 1991. In fact, Norway had leadership ambitions in international climate politics (Norwegian White Paper no. 46, 1988–1989). The Norwegian climate strategy was based on various principles including quantitative targets and timetables and flexible international arrangements. By the fall of 1991, flexibility overshadowed domestic cutbacks, and Norway was not prepared to sign an agreement that lacked joint implementation (Tenfjord, 1995). The UNFCCC resolved Norway's main concerns and the country ratified the Convention in 1993.

After Rio it became evident through the work on the national action plan that strong and far-reaching policy instruments were needed domestically to break the expected growth in GHGs by 2000. It was clear that the stabilisation target could only be reached by much tougher policy instruments than the CO_2 tax (Reitan, 1998: 145). Simultaneously, public concern for environmental protection dropped significantly in the 1990s, along with the political willingness to cope with environmental problems in general (Farsund, 1997). In 1995, after a significant delay due to conflicting interests, Norway produced a White Paper on climate policy forming the basis for reporting to the UNFCCC. Here, the country officially gave up its stabilisation goal (Norwegian White Paper no. 41, 1994–1995). In the final approach to the Kyoto Protocol, Norway focused on differential commitments and flexibility mechanisms, and proposed that Annex I countries should commit themselves to a 10–15 per cent emissions reduction by 2010. However, Norway did not want to take part in the reductions, and argued for emissions targets 5 per cent above 1990 level in 2008–2012. This number was, however, reduced to 1 per cent during the negotiations. Norway was actually among the few OECD countries that came to Kyoto without a national target (Andresen and Hals Butenschøn, 1999).

In 1997, the Labour government was replaced by a centre coalition. In 1998, this coaltion readopted a national target aiming to bring emissions back to 1989 levels by 2005. In a White Paper on climate-change policy (1997–1998), the centre coalition argued that a combination of domestic policy instruments and flexible international arrangements was needed to fulfil international commitments: 'it is neither desirable, adequate nor likely that the possibility to make use of flexible mechanisms will lead to a shift in main focus away from national measures' (St. meld. nr. 29, 1997–1998).[36] Assisted by the Kyoto Protocol, the new minority centre coalition opposed gas-fired power plants that would result in increased domestic emissions, and suggested a tax of 100 Norwegian kr. per tonne of CO_2 to cover – at a minimum – land-based industries exempt from the tax. This proposal was defeated in the Storting, and the majority decided instead to assess a future quota system allowing for free or very cheap quotas to the industry (Kasa, 1999). From 1998, the tax rates on oil and gas in the North Sea dropped significantly (Christiansen, 2000). Eventually, the centre coalition was forced to resign in 2000, partly because of its opposition to gas-fired power plants. The new Labour government subsequently signalled a shift back to a predominantly international approach. This has recently been countered by yet another new government. In 2002, the new centre-conservative government proposed the implementation of an emissions trading system in 2005. The system focuses on companies exempted from the CO_2 tax and aims at cutting domestic emissions and gaining experience before the Kyoto period from 2008 to 2012. By 2008, the Norwegian system will be expanded to all sectors as part of emissions trading under the Kyoto Protocol.[37]

Norwegian climate policy is set out in various White Papers released by different governments. In the spirit of sector integration, the Ministry of Oil and Energy also released a White Paper focusing on changing energy consumption and production in line with the Kyoto commitments (St. meld. nr. 29, 1998–1999). On paper, Norway has utilised and is planning to intensify the use of all main categories of policy instruments: regulations, economic instruments and voluntary agreements. In practice, however, Norway has relied heavily on the CO_2 tax adopted in 1991. This covers about 60 per cent of the total CO_2 emissions and some 90 per cent of CO_2 emissions on the Norwegian Shelf. It is high compared to similar taxes that have been introduced or proposed in other counties, and it is based on various tax rates for different fuels: combustion and flaring of gas in the North Sea and use of petrol have been made subject to the highest tax rates. In the eyes of the public authorities, the experience with the tax is positive. For example, a study by Statistics Norway indicates that emissions from household, transport and stationary sources may have been 3–4 per cent lower than they would have been without the CO_2 tax (Ministry of Environment, 1997).

Renewable energy (in addition to hydroelectric power) such as wave, wind, solar and bioenergy has never been high on the political agenda in Norway. In 1997, the first specific target for bioenergy and water-carried central heating was set for future market shares on new renewable energy.[38] However, targets were not set for wind and solar. In contrast to many other European countries, such as the Netherlands, incentive-based instruments to increase the use of new renewable energy have been used to only a limited degree in Norway (Christiansen, 2002).

Direct regulation based on the State Pollution Control Authority (SFT) has been used as the main policy instrument for natural gas-fired power plants. Permits granted by the SFT for natural gas-fired power plants obliged Naturkraft – owned by Norsk Hydro, Statoil and Statkraft – to slash emissions of CO_2 by half and NO_x by 90 per cent. These strict ceilings on emissions would have weakened the project's profitability based on current technology. In 1996, Naturkraft applied for a licence to build two gas-fired power stations. This application triggered a fierce environmental struggle in Norway that highlighted the tension between national and international reductions. The opponents took the position that new plants would increase total national emissions. The proponents argued that Norwegian gas could replace Danish coal and Swedish nuclear power. The Labour government supported the new plants and thought that Norwegian gas could be credited as joint implementation. Their problem was to convince the opponents that the gas would actually replace, and not be additional to, coal. The new centre coalition government combined its proposal for tax extension with opposition to gas power based on current technology. After the new Labour government took office in 2000, Naturkraft was granted permission to go ahead, but electricity prices have been too low to make any gas-fired power plants profitable.

In conclusion, Norway has until now relied almost exclusively on the CO_2 tax in its national climate policy. Additional policy instruments are mainly on the drawing board. The recently proposed national quota system aims at covering land-based companies exempted from the tax. In retrospect, climate targets and policy instruments fluctuated in their level of ambition during the 1990s, providing target groups with a highly unpredictable climate policy context. At a general level, this corresponds well with Statoil's ambiguous climate strategy.

Policy instruments and the oil industry   Energy efficiency has increased significantly in the Norwegian off-shore petroleum sector since 1990: emissions per produced oil equivalent have been reduced by 30 per cent since 1990 and those from flaring dropped by 17 per cent between 1990 and 1996. On the other hand, total emissions of GHGs have increased by some 30 per cent in the same period, and emissions are expected to increase significantly in the future. Important reasons for the expected growth are increased production and increased energy intensiveness in the production process, since such production is moving north – from the North Sea to the Norwegian Sea.

The CO_2 tax introduced in 1991 is the most important policy instrument in the Norwegian petroleum sector. While most land-based sectors are exempt from the tax, it covers almost 90 per cent of the CO_2 emissions off-shore. The tax level for burning oil and gas used to be above 300 Norwegian kr. per tonne of CO_2, before it started to drop in 1998. In 1997, Statoil launched its 'CO_2 programme' with the objective of estimating costs of implementing abatement measures on off-shore as well as land-based plants (see chapter 3). More than 50 technical options for emissions abatement on off-shore installations have been assessed, and pilot programmes were set up. According to Christiansen (2000), the CO_2 tax may have been important for decisions to implement abatement measures on off-shore installations, but taxes offer at most only a partial explanation for Statoil's CO_2 programme. Fluctuations in tax rates and recurring political discussions on replacing taxes with emissions trading have generally created uncertainty and hesitancy to invest in costly abatement measures. Since the 1970s, gas flaring has been regulated through flaring permits authorised by the Petroleum Act. While this system has led to low GHG emissions from flaring and cold ventilation compared to other countries, it has been motivated by resource-management reasons rather than climate-change considerations.

While Statoil has been somewhat influenced by the CO_2 tax to implement abatement measures, Norwegian policy on renewables has not provided Statoil with sufficiently interesting opportunities to become an energy company. Targets and measures on new renewable energy have stimulated only limited industrial development in Norway: only a few small firms are engaged in this sector and the rate of new entries in the new renewables branch is low (Christiansen, 2002). Except for some small projects on biomass and fuel cells, Statoil has mainly focused on abatement measures in its climate strategy.

In sum, the principal climate policy instrument in the Norwegian petroleum sector is the CO_2 tax introduced in 1991. This has had some impact on Statoil's climate strategy. However, Norwegian climate policy during the 1990s fluctuated in terms of targets and policy instruments. This unpredictability has made a clear proactive strategy difficult for Statoil and other target groups. In addition, public initiatives in the field of renewable energy other than hydroelectric power have to only a very limited extent provided Statoil with commercially interesting market opportunities in clean energy.

Comparison of climate policy
US climate policy has exposed its oil industry to little pressure and few market opportunities in renewables. In general, conflicting political interests sending ambiguous signals to target groups have marked US climate policy. The Clinton–Gore administration sought to develop a viable climate policy, but Congress on several occasions blocked any progress: the failure of the BTU tax as well as

insufficient funding for the public voluntary programmes are cases in point. Besides R&D, the main policy instrument in US climate policy has been genuinely voluntary programmes that have created little pressure. However, these programmes aim at creating markets for existing and close to existing technologies, thus providing industry with some opportunities. Nevertheless, ExxonMobil has largely neglected the public voluntary programmes.

In contrast to the US, there has been a broad-based political consensus on climate policy in the Netherlands. Even though there has been some political disagreement on specific reduction targets, policy instruments have been gradually stepped up over time. The 1999 climate-policy plan reduced uncertainty and shaped expectations for the future. Shell has been exposed to relatively strong regulatory pressure through the combination of regulation to improve energy efficiency and LTAs, plus the threat of rising taxes. Still, Shell's climate strategy appears to have been influenced most by ambitious targets and measures aimed at promoting market opportunities for renewable energy. ExxonMobil has been exposed to the same climate policy as Shell in the Netherlands. This observation supports the assumption that home-base countries are particularly important for corporate climate strategies.

Shell has its backbone in the Netherlands, but Shell International is actually located in the UK. While a systematic scrutiny of the UK goes beyond the scope of this chapter, we should note that the climate policy of the UK pulls in the same direction as Dutch policies, but for different reasons. The UK has reduced its CO_2 emissions and developed an ambitious climate policy due to the closure of coal mines and the shift to natural gas for economic reasons. In 2000, the UK launched its Climate Change Programme, aimed at going well beyond its Kyoto commitments to reduce GHG emissions by 12.5 per cent below 1990 levels. As part of the programme, the UK has launched an emissions trading scheme in which Shell UK and BP are major participants.[39]

Norwegian climate policy can be placed in between those of the US and the Netherlands. As in the US, there have been conflicting political interests that have led to fluctuating targets and policy instruments. This has created an ambiguous climate policy, which has made a proactive strategy difficult for Statoil.

The main difference from the US is the CO_2 tax, which has stimulated some abatement efforts. In contrast to the Netherlands, public initiatives to stimulate renewable energy in addition to hydroelectric power have not provided Statoil with sufficiently interesting market opportunities.

Linking supply and demand

People and governments, or states and societies, are linked together by institutions that channel influence. One important channel is the corporative one, in which industry, environmental organisations and governmental decision-makers meet to consult, negotiate or collaborate. Corporations do not only respond to social demand and governmental policies, they represent in themselves a social interest group with a potential to influence governmental policies. While this section will mainly focus on how political institutions affect corporate strategies, the question of how corporations exercise their influence will be more fully addressed in the next chapter.

There are two main stereotypic approaches to organising the relationship between state and industry. First, there is a conflict-oriented approach, which aims to avoid regulatory capture. Here, the state imposes standards and regulations on target groups that tend to be excluded from the decision-making process. This approach is likely to result in opposition from industry and a reactive corporate strategy. Second, there is a collaborative approach based on target-group responsibility, whereby the government consults and negotiates goals and policy instruments with target groups included in the decision-making process. This approach is likely to lead to a proactive strategy. As noted in chapter 2, it is reasonable to assume that these approaches will lead to different corporate strategies, but not necessarily different levels of environmental effectiveness. A cooperative approach is likely to lead to successful implementation at the expense of more lenient goals and policy instruments.

In this section, we will focus on the political institutions that link states and industry. Since climate change is a relatively new environmental challenge, we will include general cooperative traditions concerning environmental protection in general, as well as climate policy in particular.

The US: conflict-oriented approach

Confrontation between target groups and regulating agencies has been institutionalised in the US since the 1970s. When the EPA was established in 1970 to implement environmental legislation, Congress was concerned about 'regulatory capture', i.e. that the regulated would take control over the regulator. Legislators guarded against this by various means, which biased the EPA towards environmentalists rather than industrialists (Wallace, 1995). Concern over regulatory capture is widespread in the US, and business people are routinely excluded from regulated topics they have previously worked in.

In the mid-1980s, Vogel (1986: 21) described the US style of regulation as the most rigid and rule-oriented found in industrial society: the US makes no use of industry self-regulation, makes much use of the courts, restricts administrative discretion as much as possible and focuses on conflict in environmental policy. Ten years later, Wallace (1995: 111) still maintained that 'the adversarial, legalistic approach to environmental issues has produced an inflexible, fragmented and confused regulatory system, which stifles innovation and so frustrates industry that opposition to environmental goals seems preferable to seeking creative solutions'. The image of the US regulatory system is now changing, but the backbone of US regulatory models remains essentially intact (Dannenmaier and Cohen, 2000).

US environmental legislation is based on a number of separate acts, each focusing on a single medium: water, air and soil pollution. Each act is very detailed and leaves little flexibility with respect to implementation: *compliance* with specific standards represents the core of environmental policy. The emphasis placed on compliance by the public authorities is in turn mirrored in ExxonMobil's environmental policy (see chapter 3). While the US oil industry faces very lenient climate policy instruments, it is nevertheless subjected to strict environmental regulation in other areas of environmental degradation (water and air pollution) – perhaps even more so than in any other part of the world (Skea, 1992). Oil field exploration production is regulated at the federal level by no fewer than six laws, including the Oil Pollution Act of 1990 and the Clean Air Act as amended in 1990. In addition, federal and state regulations frequently overlap. The EPA has been developing new standards under the Clean Water Act in relation to both off-shore and on-shore exploration and production. According to the API, the petroleum industry will be spending US$26–33 billion annually to comply with the federal environmental regulations alone (Kotvis, 1994). At the other end, standards for motor vehicle design and fuel quality have a significant impact on the oil industry. For example, corporate average fuel economy standards were initiated after the energy crisis of 1974 to help reduce US reliance upon foreign oil (Kirby, 1995).

The API paints a dark picture of an industry under severe pressure, partly due to the shape and content of environmental regulations. Employment has declined significantly since 1980, profit rates have declined, US oil companies have shifted more of their activities to locations outside the US and almost half of the refineries have been closed since 1980 (Perkins, 1999). According to the Department of Energy (DoE), the oil industry spends as much on environmental protection as it spends searching for new domestic supplies of oil and natural gas. That amounts to US$10.6 billion a year, nearly twice the budget of the US EPA.[40]

Adversarial behaviour patterns are further stimulated by the court system. The threat of litigation leads to a lack of trust between regulators and the regulated, making it difficult to establish cooperative patterns. Most environmental laws have given citizens the legal right to use civil action against any person, including the administration, for failing to implement legal environmental obligations. Industry or environmental groups have on many occasions challenged the EPA. The costs of litigation sometimes even outstrip the costs of clean-up efforts (Wallace, 1995). The role of the courts in US environmental policy is also one important reason why the administration started with public voluntary programmes in climate policy: mandatory policy instruments would work against the need for swift action. For example, it took almost 10 years from the early 1980s to update the 1970 Clean Air Act. According to the API, the slowness of the US decision-making system, as well as the significant difference in how industry and government interact in Europe and the US, are very important factors explaining the difference in climate strategies between US and European oil companies.[41]

After an anti-environment stance under President Reagan in the 1980s, there was an effort to push US environmental policy in a more collaborative direction in the 1990s. For example,

President Clinton immediately established the President's Council on Sustainable Development (DSP, 1997). This council comprised 35 leaders from industry, all levels of government and environmental organisations. One important task for the Council was to develop agreement on general goals related to sustainable development. However, traditional models of interaction are not changed easily. There was generally little cooperation and consultation between the administration and industry groups before the Kyoto negotiations. In particular, the American oil industry and ExxonMobil were excluded from active participation in the decision-making process leading up to the initial US support of the Protocol. The API made an effort on several occasions to communicate with the administration. The API's perception is that the Clinton–Gore administration showed no interest in cooperating with 'Big Oil'.[42]

While access to the administration has been limited, the openness and structure of the US government provide ample room for interested parties to influence the policy process, particularly in the Congress (Skolnikoff, 1997). Fierce lobby campaigns based on funding political allies and media campaigns are a prominent part of US political culture (Kolk, 2001). Industry lobbying has also been a prominent part of energy and climate policy (Hatch, 1993). The Constitution severely restricts the freedom of action of the executive branch, that is, the administration. Like other policies, US climate policy is formulated and implemented within a political system based on the US Constitution.[43] To become law, a bill must be approved by the president and both Houses of Congress. Congress can override a presidential veto by a two-thirds vote in the House of Representatives and the Senate. Since climate change touches upon many core US interests, lobbying is the main channel for interested actors. The House of Representatives and the Senate are made up of a number of committees and subcommittees that deal with general issues. Their competence on environmental matters overlaps, and environmental legislation is dealt with by many committees. A large industry lobby exploits this fragmented structure by influencing decision-makers and assisting them with drafting legislation.

As the largest private oil company in the world, ExxonMobil plays an influential, if not a dominating, role within the API, which in turn plays an influential role in the GCC. From 1989, this organisation represented and voiced the opposition of a broad spectrum of business interests in the US, including labour, agricultural and industrial organisations. The GCC has directed and coordinated a massive opposition to US ratification from its tiny office in Washington, DC, a stone's throw away from Congress, where the coalition had a strong ally in the Republican majority of the Senate (until May 2001). Contributions to party funds provide a key channel of influence for the US fossil-fuel industries. According to Greenpeace International, the petroleum industry donated US$53.4 million to US election candidates and their political parties between 1991 and 1996 (Levy and Newell, 2000). Since 1999, ExxonMobil has been one of the largest US contributors to Republican candidates.[44]

The fate of the 1992 BTU tax is quite illustrative of the influence of US lobby organisations (Agrawala and Andresen, 1999). When the tax proposal came to the Senate Finance Committee in 1993, it was clear that it could be killed before a full vote in Senate if one Democrat voted against it. The API and a wide range of other industry interests mobilised by forming the American Energy Alliance in order to defeat the tax. The alliance hired a PR firm – Burson-Marsteller – which was able to put local pressure on Democrats through the media. Senator Boren from Oklahoma was the first to give in, thus sinking the tax proposal. According to the API, much of their resistance to the climate policy of the Clinton–Gore administration can be traced back to this event.

From 1993 to 2000, the Clinton–Gore administration tried to develop a viable climate policy at home as well as internationally by keeping 'Big Oil' at arm's length, while consulting with the green movement. This strategy proved unsuccessful because of Congressional resistance, partly as a result of intensive lobbying by the fossil-fuel industry. In 2001, George W. Bush Jr took office and the oil industry enjoyed an 'access bonanza' at the expense of consumers, according to observers.[45] As noted, ExxonMobil has been one of the most generous political donors in the US. In return, energy officials representing the Bush administration have met only with ExxonMobil and other energy industry leaders, while at the same time deliberately excluding the green movement.[46]

The US political system is to a large extent based on an adversarial approach to the development of environmental regulation.

Under the Clinton–Gore administration, the US oil industry, with
ExxonMobil in the lead, was to a large extent excluded from
participation in developing US climate policy. ExxonMobil and
the API perceived the 1992 BTU tax as extremely provocative.
This event in particular strengthened ExxonMobil's reactive
climate strategy, rather than stimulating a search for constructive
cooperation. And ExxonMobil and most of the fossil-fuel lobby
had ample room for influencing US climate policy. After the shift
in presidency in 2001, ExxonMobil and the Bush Jr administration have had complementary – if not identical – interests in a
lenient US climate policy.

The Netherlands: consensual approach
In contrast to the US, the Dutch policy system has strong neo-
corporatist qualities. The Dutch negotiated and consensus-building democracy is based on a strong state with strong social
interests. This consensual policy style has a long tradition in the
Netherlands, although its shape and content have changed over
time. For example, Arents (1999) goes so far as to claim that
consensus building is even institutionalised in the Dutch language
and that the Dutch prefer a consensual approach almost out of
habit. Even though the peak of Dutch consensual corporatism can
be traced back to the 1950s and early 1960s, negotiated agreements with industry during the 1990s still represent visible signs
of this tradition.

Bargaining and cooperation with interest groups, both
formally and informally, are particularly evident in the environmental sector, including climate policy. The Netherlands has
developed a target-group approach based on the idea of raising
environmental concerns and social responsibility among those
actors causing the problems in the first place. In return, the
government lets the target groups have their say in the making
and implementation of environmental policy and pays serious
attention to their needs.

The use of negotiated agreements in the Netherlands is part of
a comprehensive environmental policy in which target groups
have actively participated all the way. The Ministry of
Environment started developing a consensual approach on the
basis of the Environmental Protection Act of 1980. In 1981, the
Council for the Environment was established, providing environ-

mental organisations, industry and other actors with a formal channel for influencing environmental policy (Liefferink, 1995). The ministries' cooperative approach was further developed in the indicative multi-year programmes for the environment, which started in 1984. The way in which NEPP 1 was developed represents a deepening and a continuation of the target-group approach. In a relatively short time, this approach led to approximately 100 negotiated agreements, or covenants as the Dutch call them, covering all major industrial sectors. NEPP 1 was supposed to be published in 1987, but was delayed for two years owing to the ambition of seeking agreement with all parties affected by specific goals and policy instruments (Bennett, 1991).

Shortly after the adoption of NEPP 1, the Committee on Environment and Industry was established by the major actors representing the authorities and industry (Suurland, 1994). Fourteen industrial sectors were selected as priority target groups, involving some 12,000 companies responsible for over 90 per cent of industrial pollution. On the basis of the NEPP targets, the Ministry of Economic Affairs developed a programme aiming to increase energy efficiency in industry by 20 per cent by the year 2000. LTAs are signed by the sector association, individual firms and the Minister of Economic Affairs. All agreements include an objective, energy-conservation strategy, energy-saving plans for individual firms, monitoring and statement of duration.

Negotiations between the government and industry – particularly the Netherlands Employers' Association (VNO) – produced a broad-based consensus on ambitious targets, including the CO_2 stabilisation target and the need for increasing energy efficiency by 20 per cent between 1990 and 2000. This agreement was based on differentiated targets between companies within the same branches. In homogeneous branches, agreements were concluded between the branch organisations and the government in order to reduce costs. As could be expected, the involvement of target groups forced the government to agree to less stringent goals than it otherwise would have proposed (Weale, 1992).

In contrast to the US, the Dutch oil and gas industry has traditionally operated with a high level of discretion concerning both the identification of problems and the implementation of measures (MILJØSOK, 1996). Accordingly, the Dutch oil industry has a more positive experience with environmental regulation. Air pollution from off-shore activities has been weakly regulated compared to on-shore ones, and the authorities have had few sanctioning opportunities in cases of non-compliance. Standards and technology requirements have been negotiated within the framework of the permit system.

The Dutch consensual tradition and target-group approach have affected the way Shell operates as a company.[47] Until 1995, the Shell Group was based on an unusually complex organisational structure. Complexity was further increased by the Group's emphasis on reaching decisions by *consensus*, which implied that decision-making involved an unusually high level of internal discussion (see chapter 4). In essence, Shell mirrored the Dutch consociate democracy. The Dutch way of organising contact between industry and government has also led to close cooperation between Shell Netherlands and the authorities. According to Shell, the company has a very good relationship with the Dutch government on environmental matters.[48] Cooperation between Shell and the authorities takes place in formal committees as well as informally. This relationship does not, however, imply that Shell and the government have identical interests in all matters. Shell and other Dutch companies have had their disagreements on climate policy, particularly related to the regulatory tax. Like most other companies, Shell dislikes environmental taxes.

In 1989, there was a change in the Dutch government leading to a revision of the first NEPP. The new plan – NEPP Plus – disturbed the original platform of consensus on which the first NEPP was based. NEPP Plus introduced a more stringent target and emphasised economic instruments. The plan was criticised both for going too far and for not going far enough. Industrial and agricultural target groups warned about moving too far ahead of other EU countries. On the other hand, the environmental movement thought that the new goals and instruments were insufficient to establish sustainable development. The new CO_2 reduction target and the emphasis on economic policy instruments had not been agreed upon with industry, including the oil industry. Industry ferociously resisted the upcoming tax, and the VNO withdrew its formal support for NEPP Plus (Wallace, 1995). While this setback may have had consequences for the credibility of the climate target, it had no negative consequences for the LTAs adopted in the field of energy efficiency. Since 1991,

29 agreements have been signed, representing about 90 per cent of total energy use in industry, and the 20 per cent energy efficiency target is in sight (Skjærseth, 2000).

While employer, car and oil associations supported the energy-efficiency targets, they continued their fight against the regulatory tax, arguing that financial instruments should be introduced only at the EU level. At the same time, the same interest groups lobbied intensively against the EU-proposed carbon/energy tax through their respective Eurofederations: the EU tax proposal eventually failed in the Council of Ministers shortly before the 1992 Rio conference. The European Petroleum Industry Association (EUROPIA) expressed the strongest reservations concerning the creation of a tax on oil products. Shell was a key actor at both the Dutch and the EU levels and pointed to a contradiction between SO_2 and CO_2 reduction targets. In order to reduce one tonne of SO_2 emissions, refineries were forced to produce 10 additional tonnes of CO_2 emissions (Schenkel, 1998). This argument is still being used by EUROPIA, even though it has shifted to a more positive stand on the need to combat climate change.[49] In the end, the oil industry and other interests won a partial victory. On 1 January 1996, a combined carbon/energy tax was introduced in the Netherlands unilaterally. It is levied on the use of natural gas, heavy fuel oil, diesel oil, liquified petroleum gas and electricity. Natural gas and electricity produced from renewable sources are exempt. On the other hand, the biggest industrial users pay only a third of the price that small users – households and small and medium-sized firms – have to pay.

In sum, the Netherlands has, in contrast to the US, a consensual tradition, and it has based its environmental and climate policy on close cooperation between target groups and the government. This has affected the corporate structure of Shell and has led to good relations between Shell and the Dutch government, which in turn has stimulated a proactive strategy. However, like most other corporations, Shell opposed the regulatory tax.

Norway: a mixed approach
Like the Netherlands, Norway is frequently classified as a neo-corporatist country where institutionalised rights of participation are provided to non-governmental organisations in all phases of governmental policy. However, Norwegian environmental policy

has been based neither on extensive formal target-group participation, nor on exclusion. In contrast to the adversarial US
approach, the Norwegian counterpart to the EPA, the SFT, has
traditionally given priority to industry in its licensing policy,
although priority given to environmental protection interests has
increased (Gjerde, 1992). In contrast to the comprehensive and
inclusive Dutch environmental programmes, formal channels for
participation in Norway are less developed and more ad hoc. In
addition, there is general scepticism towards the use of voluntary
agreements in Norway. They have never made their breakthrough
in Norwegian climate policy, in contrast to many other OECD
countries.[50]

Statoil is partly owned by the Norwegian state and has traditionally served as a state instrument for protecting Norwegian
petroleum interests. The link between Statoil and the state is thus
particularly strong, and Statoil has a privileged status as a 'core'
insider. However, Statoil has increasingly been treated on equal
terms with other oil companies by the licensing system that regu-
lates petroleum operations in Norway (Rudsar, 1999). And, as we
will see, the relationship between the government and Statoil in
climate policy can be characterised by both conflict and cooperation.

The Petroleum Act and the Pollution Control Act regulate
emissions other than those of GHGs to the sea and air. Norway's
Petroleum Act requires environmental impact assessment to be
carried out at several stages in petroleum operations – from the
opening of an area to the disposing of abandoned installations.
On the basis of such assessments, the Storting undertakes an
assessment of the pros and cons of pursuing operations in the
area. Discharges to the sea are regulated by the Pollution Control
Act in the form of individual licences. The principal rule is zero
discharges of hazardous substances, in line with the 1995 North
Sea Declaration. However, the Norwegian off-shore regulatory
regime is generally described as flexible in practice: standards are
general and leave significant scope for interpretation on a case-
by-case basis. Sanctions are rare and compliance is mainly based
on internal control conducted by the companies themselves
(MILJØSOK, 1996:100). As in the Netherlands, the oil industry
appears generally satisfied with 'traditional' environmental regulation.

When the Norwegian climate target was discussed in the parliament in 1989, industry did not show much interest in the issue, and the Norwegian Employers' Association (NHO) did not have any strong opinions (Bolstad, 1993). In the same year, the Environmental Tax Committee was established to assess the foundation for a CO_2 tax. This committee was expert-dominated and did not include representatives from interest organisations. Moreover, the first report, which provided the basis for the 1991 tax, was not made public. Thus, industry and environmental organisations were effectively prevented from formally influencing the premises for the tax. And the parliamentary decision to implement the tax mainly on off-shore activities was made in the face of strong resistance from the entire petroleum industry (Reitan, 1998; Kasa, 2000). However, the strongest resistance came from energy-intensive land-based industry fearing an expansion of the tax. This included Statoil and another large government-owned company with a big petroleum division, Norsk Hydro. These companies argued that neither gas-fired power plants nor planned methanol plants would be economically viable with a CO_2 tax. Nevertheless, the most important economic sector for Norway and the Norwegian oil industry – off-shore oil and gas production – lost in its opposition to the CO_2 tax, while the energy-intensive land-based industries won.

In 1994, the Green Tax Commission was established to assess a transformation from taxes on labour to taxes on pollution. Extension of the CO_2 tax to land-based industries rapidly became one core topic within the commission. In contrast to the Environmental Tax Committee, the Green Tax Commission was based on interest representation, including the Naturvernforbundet and Norsk Hydro. The Naturvernforbund, the Ministry of Environment and the Ministry of Finance preferred a comprehensive tax based on environmental cost-effectiveness concerns. The industry, represented by NHO, the largest labour union (Landsorganisasjonen) and the Ministry for Trade and Industries (in which the previous Ministry of Oil and Energy was included), was more concerned about cost distribution than cost-effectiveness. It argued that exports and employment would decline particularly in small rural towns heavily dependent on energy- intensive industries. In the end, prime minister Gro Harlem Brundtland (Labour) decided to block an extension of the tax. One important reason for her decision was the planned gas power plants (Kasa and Malvik, 2000).

In 1997, a new centre government replaced Labour. The new government opposed gas power, and favoured an extension of the CO_2 tax. The Ministry of Trade and Industry and the Ministry of Petroleum and Energy were split and largely decoupled from their traditionally close relationship with the industry under the Labour government. However, industry interests changed their strategy and lobbied towards the political parties in the Storting. Eventually, a majority voted against extending the tax and decided to prioritise an emission trading system (Kasa and Malvik, 2000). In short, Statoil was to a large extent formally excluded from the Norwegian CO_2 tax process, and the CO_2 tax was adopted and implemented against the interests of Statoil.

After the first round of wrangling about the CO_2 tax, Statoil and the Ministry of Petroleum and Energy initiated in 1995 a new cooperative forum, MILJØSOK, in order to improve environmental performance on the Norwegian continental shelf. This forum aimed to improve cooperation primarily between the oil industry and the authorities in the field of the environment based on the NORSOK model. NORSOK was established in 1993 to improve the competitive standing of the Norwegian offshore sector. MILJØSOK represented a voluntary approach in which Statoil held a key position through its leadership of the steering group (MILJØSOK, 1996). The objective of MILJØSOK was to contribute to a more effective environmental strategy as well as improve cooperation between the authorities and the industry. It operated as a 'signpost' which reduced uncertainty and stimulated innovative solutions (Christiansen, 2000). A final report from phase one concluded that tax rates should be reduced in favour of establishing binding forms of cooperation, such as negotiated agreements. Transition to cleaner technologies was identified as the most cost-effective approach to problem solving in this sector.

MILJØSOK initiated a second phase in 1997 (MILJØSOK, 2000). In this phase, ambition levels were significantly reduced, and it became evident that radical measures were needed to counterbalance expected increases in emissions owing to higher production levels and higher energy needs. Nevertheless, according to the head of Statoil's CO_2 programme, the MILJØSOK initiative played an equally if not more important role than the CO_2 tax for Statoil's climate strategy. Decisions on more energy-efficient power-generation technology have been strongly influenced by the findings of the MILJØSOK initiative (Christiansen, 2000).

An effort was also made to reduce emissions of non-methane volatile organic compounds (NMVOC) from shuttle tankers. The formation of ground-level ozone is, however, the main problem associated with NMVOC in combination with NO_x. Negotiations began in 1998 between the Norwegian authorities and the oil industry on an agreement for reducing NMVOC emissions from shuttle tankers loading crude oil. Such activities account for about 60 per cent of total NMVOC emissions in Norway. The parties comprised the Ministry of Environment plus the oil industry, which in turn was composed of 18 companies with licence interest on the Norwegian continental shelf. The aim of the agreement was to apply best available technology (BAT) on all 20 relevant ships by 2005. In total, this would cost about 2 billion Norwegian kr. and lead to reductions of the order of 70 per cent from each ship. Introduction of BAT on the shuttle tankers is considered to be the most effective means in terms of both costs and effects in the petroleum sector (Dragsund et al., 1999). In the end, however, major US oil companies suddenly refused to support the deal and the agreement collapsed. This example shows that multinational companies can affect governmental policies of host countries as well as home-base countries. Moreover, it points to the link between policy instruments and the climate strategies of multinational target groups (see chapter 7).

Norway can be placed between the US and the Netherlands with respect to political institutions regulating state–industry relationships. In climate policy, the foundation for the CO_2 tax was shaped without formal participation by Statoil and adopted in spite of Statoil's interests. On the other hand, Statoil has been the key oil company participant in the MILJØSOK initiative, which aimed to improve cooperation between companies and the authorities. According to Statoil, this cooperative forum has proved as important for the company's climate strategy as has the tax initiative.

Conclusion

To recapitulate, the DP model was based on the assumption that even multinational companies are particularly influenced by their home-base countries. Companies closely tied to home-base countries with a high social demand for environmental quality, a governmental supply of ambitious climate policy, and political institutions that promote cooperation and consensus seeking with target groups tend to adopt a more proactive strategy. Conversely, companies with home-base countries characterised by a relatively low social demand, weak climate policy and political institutions based on conflict and imposition tend to adopt a more reactive response. The results are presented in table 5.3.

Table 5.3 *The DP model: expected versus actual strategies in relative terms*

Home country	Social demand	Governmental supply	Political institutions	Predicted strategy	Actual strategy
The Netherlands	High/ Medium	High	Consensus	Proactive	Proactive
Norway	Medium/ Low	Medium	Mixed	Intermediate	Intermediate
US	Low	Low	Conflict	Reactive	Reactive

Table 5.3 shows that the propositions derived from the DP model gain strong empirical support evaluated in terms of pattern-matching: all three cases roughly match the expectations derived from the DP model. Differences in national political context vary systematically with differences in corporate climate strategies. Variation in social demand, governmental supply and political institutions linking demand and supply all apparently matter in explaining the climate strategies of ExxonMobil, Shell and Statoil. Judging the DP model in terms of the tenability of the propositions derived from it, we can conclude that we have had a high degree of success in predicting corporate climate strategies.

Correlation, however, is not the same as causation. The causal patterns linking domestic political context to corporate strategy appear long and even indirect. Below, we will first recapitulate the main relationships between the DP model and corporate strategies before causal patterns and mechanisms are critically examined.

The relationships between national political context and corporate climate strategy appear strong. Shell has its roots in the Netherlands and has adopted a proactive climate strategy. The Dutch population expresses the highest demand for climate policy, the Netherlands has adopted the most ambitious climate policy, and the political institutions are based on cooperation and consensus-seeking between the state and industry. This pattern appears robust even if we control for the fact that Shell International is located in the UK. In contrast, the North American company ExxonMobil has adopted a reactive climate strategy within a significantly different political context. North Americans apparently express less concern about climate change, US climate policy is weak in terms of goals and policy instruments, and political institutions channelling state–industry influence can be characterised as adversarial. In between, we find Statoil and Norway. Statoil has adopted an ambiguous climate strategy which can be characterised as intermediate, i.e. neither reactive nor proactive. The Norwegian population has fluctuated significantly in its concern for climate change, and so has Norwegian climate policy. In addition, political institutions linking state and society are mixed, i.e. based either on close cooperation or on conflict. Table 5.3 also depicts a strong relationship between social demand and governmental supply of climate policy. While social demand matters for climate policy, other factors such as energy-economic circumstances are probably equally important. Nevertheless, social demand appears to affect corporate strategy both directly through consumer behaviour and indirectly by influencing public policy.

At the level of general patterns, we have thus no empirical observations supporting the assumption that state-owned companies are more likely than private companies to choose a climate strategy in accordance with the position of their governmental owners. The relationship between home-country context and climate strategies appears to be strong independent of type of ownership. This observation is also supported by the fact that the Norwegian CO_2 tax was introduced in spite of opposition from the state-owned company representing the strongest economic sector in Norway: Statoil. Looking also at the introduction of the Dutch regulatory tax against the will of Shell and other Dutch energy-intensive industries, we see that the state can maintain a certain degree of independence from business interests in climate policy. The Clinton–Gore administration also made an effort to distance itself from ExxonMobil and US fossil-fuel industries by proposing the BTU tax. In this case, however, the government was defeated by industry lobbying directed at Congress.

Below, we will take a critical look at the causal mechanisms and patterns linking the political context of the companies' home-bases to corporate climate strategies. Let us start with the observations supporting a close link between the Netherlands and Shell. On the one hand, we have strong indications that Shell has been particularly sensitive to the Dutch societal and political context. The company has used the Netherlands as a sort of test case for predicting changes in energy carriers in the future. A relatively high social demand is in line with Shell's scenarios and probably linked to its strategy on renewables. However, Shell's vulnerability to consumer campaigns and loss of reputation must be understood in the context of the company's experiences with boycotts linked to South Africa, Nigeria and Brent Spar, which have increased the company's sensitivity to societal pressure. Accordingly, Shell's climate strategy along this dimension should be looked at in the light of the interface between political context and corporate specific events (see chapter 4). Dutch supply of climate policy in the 1990s was characterised by relatively ambitious targets and the adoption of increasingly vigorous policy instruments. Regulation and negotiated agreements on energy efficiency backed up by threats of taxes put some pressure on Shell. However, Dutch policies on renewables appear more directly linked to the company's climate strategy. An ambitious governmental policy on renewables combined with a social demand for clean energy provided Shell with an anticipation of new market opportunities. Cooperative political institutions also contributed to constructive relationships between the Dutch authorities and Shell. In essence, the link between Dutch national political context and Shell's proactive climate strategy is marked by a positive interplay between pressures and opportunities provided by social demand, governmental supply and political institutions linking demand and supply. Moreover, it is marked by a positive interplay between political context and company-specific factors.

On the other hand, this pattern of observation does not appear sufficiently persuasive. Although Shell has its roots in the Netherlands, the country represents only a part of the company's operations and activities. Shell operates in more than 135 countries, and the main markets are located in Europe and the US. Relying on developments in the Netherlands alone thus appears too risky. In addition, the DP model as applied to the Netherlands does not account sufficiently well for the *changes* witnessed in climate strategy. Shell changed from a reactive to proactive company around 1997–1998 and was a member of the strongest and most aggressive US-based anti-climate lobby group – the GCC – until 1998. Social demand peaked around the Rio conference in 1992 and then declined somewhat. Dutch climate policy progressed steadily in terms of targets and policy instruments during the 1990s, without any dramatic changes. The question then becomes whether the Dutch target on renewables adopted in 1995 and the Dutch LTA on energy efficiency concluded in 1996 created sufficient pressure and opportunities to trigger a turnabout in Shell's climate strategy. Although it is difficult to judge, we believe that these changes in policy were important, but not sufficient to push and pull a large multinational oil company towards a more proactive strategy.

Turning to the US and ExxonMobil, we find some answers but also some new questions. In the 1990s, the relationship between ExxonMobil and the US comes close to the opposite of the relationship between Shell and the Netherlands. Social demand in the US has not created any pressure or provided sufficiently interesting market opportunities to ExxonMobil. Public concern for climate change has been relatively low and the environmental movement has not placed any strong pressure on the climate strategy of ExxonMobil in the US. This may also be related to a widespread perception of ExxonMobil as a 'super-tanker', insensitive to public pressure due to its size and power. Equally important is the North Americans' reluctance to pay a higher price for clean energy. Likewise, US supply of climate policy has been marked by lack of pressure and opportunities. US climate policy has been based on public voluntary programmes in which the industry has been invited to participate. These programmes aim to stimulate markets for energy-effective technologies, but ExxonMobil has not paid much attention to them. Genuinely voluntary programmes are most likely to work in a different social context marked by a higher social demand, such as the Netherlands.

The US political institutions linking state and society have traditionally been adversarial. ExxonMobil has brought this tradition further, as one of the leaders of the US fossil-fuel lobby, by lobbying forcefully and apparently successfully against the climate initiatives of the Clinton–Gore administration from 1993 to 2000. The relationship between the oil industry and US government has in this period been marked more by a state of Cold War than by constructive cooperation. In short, the link between the US and ExxonMobil's reactive climate strategy is marked by a negative interplay between pressures and market opportunities. Social demand for climate policy has been weak, governmental supply of climate policy has been weak, and the US political system is based on an adversarial approach that has stimulated a reactive strategy against the initiatives taken by the US administration in the 1990s. According to the DP model, the climate strategy of ExxonMobil can be understood as a combination of various domestic factors pointing in the same direction. In addition, the US context is marked by an interplay between political context and company-specific factors: ExxonMobil was in a way predisposed to a reactive strategy to a more significant extent than Shell and Statoil (see chapter 4).

This being said, ExxonMobil's strategy also leaves us with uneasiness against the backdrop of the DP model. The main cause for worry is again related to change: ExxonMobil did not show any signs of changing its climate strategy between 1997 and 2001. In this period, the US had signed the Kyoto Protocol demanding a dramatic change in US climate policy. The principal US architect behind the deal struck in Kyoto was Vice-President Al Gore, running against Bush for presidency in the autumn of 2000. While ExxonMobil and the rest of the US fossil-fuel lobby controlled domestic climate initiatives, the international process exposed ExxonMobil to significant uncertainty over this four-year period. In addition, the European market is also important to ExxonMobil (see chapter 3). A likely response would thus be to prepare for the worst case – a Kyoto Protocol in force including the US – but there is no evidence that ExxonMobil had a 'plan B' in the case of US ratification.

The relationship between Statoil and Norway appears more

clear cut. The reason is not that Statoil is partly owned by the Norwegian state, but rather the high degree of overlap between Norway and the Scandinavian countries and Statoil's main markets. In Norway, social demand for climate policy has fluctuated significantly over time, thus generating quite ambiguous market signals. Statoil has been exposed to some pressure from ENGOs, but has not experienced large-scale consumer campaigns linked to any aspect of the company's activities. Norwegian climate policy has also been ambiguous. Targets on GHG emissions have been adopted, abolished and readopted, and Norwegian policy on renewables has not provided any strong incentives for Statoil. On the other hand, Norway was one of the first countries to adopt a CO_2 tax, which applied specifically to the Norwegian continental shelf. This tax influenced the realisation of Statoil's CO_2 programme, which aimed to increase energy efficiency. Statoil played the key role in the MILJØSOK initiative aimed at identifying solutions and improving environmental cooperation between the state and oil companies. MILJØSOK has also been seen as an important explanation for Statoil's climate strategy. In short, Statoil's ambiguous climate strategy appears closely linked to an ambiguous Norwegian climate policy context.

This chapter has provided some additional answers, but also raised new questions – particularly related to the conditions triggering change in the climate strategies of the oil companies. Why did ExxonMobil not modify its reactive strategy after the US consent to the Kyoto Protocol? And was Shell's turnabout caused exclusively by changes in the Dutch political context? In the next chapter, we explore these questions by analysing developments on the international level. We thus move beyond the analysis of single companies within single home countries to improve our understanding of *changes* in corporate climate strategies in particular. Our main assumption is that additional answers can be found at the interface between corporate influence on international regimes and regime influence on corporate strategies.

Notes

1 'Esso says effect of UK protests not yet clear'. Source: www.planetark.org (accessed 5 December 2001).
2 Source: www.statoil.com (accessed 28 March 2001).

3 Personal communication with Barend van Engelenburg, Ministry of Housing, Spatial Planning and the Environment, 28 November 2000.

4 See Rowlands, 2000, for comparisons of national attitudes to climate change.

5 Results from 39 different reports focusing on this issue have been collected and analysed by the Ministry of the Environment. Twice a year since 1980 the research agency NSS/market onderzoek has measured the level of involvement among the Dutch population with regard to some 40 social issues.

6 It is important to emphasise that there are a number of methodological problems related to the measurement of citizens' willingness to pay. The results obtained from such analyses may depart significantly from actual consumer choice.

7 However, environmental problems related to air, water and soil rank higher in public awareness than do energy issues and climate change.

8 Personal comunication with Ir. Henk J. van Wouw, Shell Netherlands, 29 November 2000, and Barend van Engelenburg. Ministry of Housing, Spatial Planning and the Environment, 28 November 2000.

9 This study is based on lead news stories on 'global warming' in the *New York Times* and the *Washington Post.*

10 Personal communication with Brian P. Flannery and Gary F. Ehlig, ExxonMobil, Irving, Texas, March 2000.

11 Personal communication with Iain MacGill, senior policy analyst, Greenpeace Climate Campaign, Greenpeace, Washington, DC, 23 March 2000.

12 The MORI poll sampled 998 adults in 191 places across Great Britain. The results show that the share of customers that regularly bought petrol from ExxonMobil subsidiary Esso was reduced from 26 per cent in August 2001 to 19 per cent in July 2002. The share of customers that bought their petrol from Esso's rival retailer, BP, rose from 18 per cent in 2001 to 21 per cent in 2002. According to the head of environmental research at MORI, the change was 'statistically significant, beyond the margins of error'. (Source: www. planetark.org (accessed 5 September 2002).

13 Election data are not very robust, since environmental attitudes connected with elections are closely linked to the political context of each election.

14 Norsk Monitor is conducted by the Norwegian polling agency MMI (Markeds og Mediainstituttet). The data are an extension of those used in Norsk Monitor, based on communication with Ottar

Hellevik, Director of MMI.

15 *Stavanger Aftenblad*, 28 January 2000, 'Statoil trekker seg fra vind-kraftprosjekt' ('Statoil withdraws from windpower project'). Source: http://aftenbladet.no/nyheter/okonomi/article.jhtml?articleD=5730 (accessed 19 September 2002).

16 *Aftenposten*, 1 December 2001, 'Natur og Ungdom Anmelder Esso' ('Natur og Ungdom [the youth organisation of Naturvernforbund] starts lawsuit agaisnt Esso').

17 A study of the causes of climate policy would require an in-depth analysis of factors such as energy-economic circumstances, change in government and the position of the ruling party, governments' control over the legislature, distribution of competence between regulatory agencies, and different regulatory styles embedded in national history and tradition.

18 To enter into force at least 55 countries will have to ratify the Kyoto Protocol. In addition, the protocol has to be ratified by Annex I countries accounting for at least 55 per cent of the total CO_2 emissions from Annex I countries in 1990.

19 The climate policy of the US was in this period developed in two documents. First, 'America's climate change strategy: an action agenda' was presented at the first INC session. Second, 'The US national climate change action plan' was presented at the sixth INSC meeting.

20 Other key elements underlying the US position were: (1) reluctance about additional financial support to developing countries to meet their obligations; and (2) ambition to create a strong but flexible international regime.

21 These concepts are borrowed from research primarily focusing on implementation of national policy. However, they are also applicable to joint international commitments. For example, one factor that distinguishes between them is whether the initial focus is a central governmental decision (top-down) or a network involved in a policy area (bottom-up). See e.g. Sabatier, 1986.

22 For example, in 1996 and 1997, only 60 per cent of the funding requested by the president was approved by Congress (Brunner and Klein, 1998).

23 Emissions increased by 14.5 per cent between 1990 and 2000. Adding the 7 per cent reduction Kyoto target, the US would have to reduce emission by 21.5 per cent before 2012 even in the absence of the expected annual increase in emissions.

24 'President announces clear skies and global climate change initiatives. Source: www.whitehouse.gov/news/releases/2002/02/20020214-5.html (accessed 15 February 2002).

25 See DSP, 1997, for a description of the various programmes.

26 Personal communication with Brian P. Flannery and Gary F. Ehlig, ExxonMobil, Irving, Texas, March 2000.

27 This EPA-run programme seeks to encourage natural gas companies to adopt cost-effective technologies that reduce emissions of methane. The programme includes transmission, distribution and production (the last after 1995). The EPA provides implementation support, public recognition and removal of unjustified regulatory barriers, and the programme has identified various methane-reducing best-management practices. Participating companies submit a plan to the EPA, and implement it over the next three years.

28 Source: www.epa.gov/gasstar/annual.htm (accessed 29 October 1999).

29 Climate change may have striking consequences for the one-third of the Netherlands lying below sea level.

30 Also in 1990, the Rijkswaterstaat published the 'Impact of sea level rise on society' report. This concluded that the sea level may rise 60 cm per 100 years and the need for measures would become urgent from 2000 (Schenkel, 1998).

31 Fiscal incentives are provided by a number of provisions in corporate income tax: energy conservation and renewable energy are encouraged by these, and a tax credit was introduced in 1996.

32 Personal communication from Ir. Henk J. van Wouw, manager of environmental affairs, Shell Nederland BV, 28 November 2000.

33 The Dutch strategy was developed before the presidency period, and rested on two main principles. First, a sector approach was followed rather than a country-by-country one in calculating reduction potential. Second, the total EU target became directly linked to the size of national contributions to this target.

34 While the Kyoto protocol includes six GHGs, the EU agreement covers three substances: CO_2, methane and N_2O.

35 This target was made conditional upon international and technological developments as well as further research.

36 Note that even though the reduction of domestic emissions is given the highest priority, the proposed climate package is expected only to cause emissions reductions of about half the expected rise of CO_2 emissions.

37 'Norway to start emissions quota trading in 2005'. Source: www.planetark.org (accessed 25 March 2002).

38 Utilisation of bioenergy and water-carried central heating should be increased by 5 TWh over a period of 5–10 years, representing about 5 per cent of total power production.

39 '"Hot air" blows gaping hole in the emissions trading scheme',

ENDS Report, 326: March 2002.

40 Environmental Research and Analysis, US Department of Energy – Office of Fossil Energy. Source: www.fe.doe.gov/oil_gas/oilgas7.html (accessed 15 April 1999).

41 Personal communication with Phil Cooney and Bill O'Keefe, API, 21 March 2000.

42 Personal communication with Phil Cooney and Bill O'Keefe, API, 21 March 2000.

43 The US Constitution was designed to demarcate power between the three branches of the federal government: the legislature, composed of the House of Representatives and the Senate, which together make up the Congress; the executive branch, including the president and federal agencies and departments; and the judiciary, consisting of federal courts.

44 *New York Times*, 27 March 2002, 'Documents show energy officials met only with industry leaders'.

45 'Bush energy plan said to help industry, not public'. Source: www.planet.ark.org (accessed 24 January 2002).

46 *New York Times*, 27 March 2002, 'Documents show energy officials met only with industry leaders'.

47 Personal communication with Barend van Engelenburg. Ministry of Housing, Spatial Planning and the Environment, 28 November 2000.

48 Personal communication with Ir. Henk J. van Wouw, manager of environmental affairs, Shell Netherlands, 29 November 2000.

49 Personal communication with Valèrie Callaud, EUROPIA, 30 November 2000.

50 Voluntary agreements have mainly been used in order to combat GHG emissions from the aluminium industry.

6

The International Regime model

In the preceding chapters, we have analysed the climate strategy choices of the oil industry as a function of company-specific factors (the CA model analysed in chapter 4) and of factors linked to the domestic political context in the home-base countries of the companies (the DP model analysed in chapter 5). These models have provided us with some answers as to why the climate strategies of the oil companies differ, but have left other questions unanswered. In particular, we do not have a good understanding of Shell's turnabout from a reactive to a proactive company. Additionally, it is difficult to understand on the basis of the CA and DP models why ExxonMobil did not modify its reactive strategy in the four-year period from the US's signing of the Kyoto Protocol until Bush Jr was elected president.

In this chapter, we shift our focus from the domestic to the international level. To what extent can the international climate regime explain the strategies chosen by the oil industry? Climate change is a global problem partly caused by global actors operating within the framework of international institutions. It is thus natural to move beyond the study of single companies within single states to get a comprehensive understanding of corporate climate strategies. Accordingly, the International Regime (IR) model is concerned with corporate alliances across states and how such alliances relate to international regimes over time.

Recalling our propositions discussed in chapter 2, we assume that corporate climate strategy is likely to be formed through the two-way influence between regimes and industry. This approach permits a stronger focus on *changes* in corporate strategies over time. The argument is straightforward: if industry determines joint international climate commitments, the regime is unlikely to affect industry strategy. Conversely, if industry exerts little influence over joint international climate commitments, the regime is likely to affect industry strategies. More specifically, we assume that to the extent that a reactive industry exerts a *strong influence* on the international climate regime, a reactive strategy is likely to persist. The international climate regime provides industry with the opportunity of exercising influence far beyond single home countries. The regime channel is particularly important for large multinational companies operating all over the world. Our analysis of the influence exerted by a reactive industry on joint commitments in the climate regime is based on three assumptions. First, we assume that the influence of target groups tends to increase the more cohesive their strategies are. Second, influence is likely to depend on the strength of counterbalancing forces such as ENGOs. Third, we assume that influence will increase the more access industry has to decision-making processes and the fewer actors need to be persuaded to block collective decisions. The last point is related to the decision procedures applied by the regime.

To the extent that a reactive industry exerts a *weak influence* on the international climate regime, we assume that the industry is more likely to shift towards a more a *proactive* climate strategy as a response to regime development. In contrast to the policy of home-base countries, a progressing international regime carries the potential of affecting multinational companies all over the world. The impact of the climate regime on strategy choice is analysed in terms of three causal pathways: knowledge, pressure and opportunities.

From Rio to Kyoto: corporate influence on the climate regime

It is difficult to separate the influence of one corporation in particular at the international level since companies tend to coordinate their positions within industry organisations to match the scope of the international regimes. We will thus explore the influence of the fossil-fuel lobby in general, with a particular focus on ExxonMobil and Shell. To measure the influence of the fossil-fuel lobby, we proceed in two steps. First, we deploy the criterion of goal attainment to analyse the match between industry positions and joint international commitments. Goal attainment in this context refers to the extent to which the industry succeeded in achieving its goal of preventing ambitious GHG emissions reduction targets from entering the joint international climate commitments. We focus on the UNFCCC, the Kyoto Protocol, and the post-Kyoto process leading up to the US withdrawal from the Kyoto agreement. Second, we search for observable influence within the two most important actors that to a large extent have determined the international commitments: the US and the EU. Notice that focus on the US is changed from domestic climate *policy* (chapter 5) to the development of *positions* in international negotiations. It is also important to note the dual role of the EU in this context. On the one hand, the EU is an actor within the framework of the international climate regime. On the other hand, the EU constitutes in itself a regime, with an increasingly more ambitious approach to the climate problem. While the EU generally is clearly different from international regimes with regard to its scope, depth, nature and competence, EU environmental policies in specific issue areas, such as climate change, face many of the same implementation challenges as international regimes (Skjærseth and Wettestad, 2002).

Corporate influence on the UNFCCC
Unity, mismatch and divergence characterise the past decade of the relationship between the climate regime and fossil-fuel industry interests. In the prelude to the UNFCCC, the fossil-fuel lobby, including ExxonMobil and Shell, was unified in its opposition to binding and specific international climate targets. All major oil companies took the position that action on global warming could be damaging to their economic interests. Moreover, the fossil-fuel lobby had a near monopoly over the entire business and industry perspective:

> In the early 1990s, the idea of significantly limiting greenhouse gas emissions would have been opposed by almost all the major industries on the planet: the fossil fuel industries, 'from production to the pump', along with most manufacturing, processing and automobile industries. The rest of the corporate world would at best have been indifferent, with the almost trivial exception of the nascent energy efficiency and non-fossil energy industries. (Grubb et al., 1999: 257)

Industry did not necessarily reject the possibility of an emerging climate-change problem, but industry organisations emphasised 'no regrets' measures (measures that were already justified on other grounds), national flexibility and voluntary measures. Moreover, industry placed emphasis on the global nature of the problem and problem solving 'abroad' in less developed countries and those with economies in transition. This focus would distract attention from domestic regulation. For example, the Union of Industrial and Employers' Confederation of Europe (UNICE) emphasised assistance to eastern European countries as a core strategy towards combating climate change (UNICE, 1991).

Until 1992, the fossil-fuel industry, particularly in the US, had a good grip on the international process. The UNFCCC was roughly in line with the interests of the fossil-fuel lobby and the oil industry. The API, for instance, was pleased with the outcome.[1] It did not include any binding 'targets and timetables' for emissions reductions, and it did not restrict the parties' choice of policy instruments. In essence, the Convention did not exert pressure or create significant market opportunities for industry. It did, however, place the responsibility for the climate problem and its solution on developed countries, and established an institutional framework for reporting action as well as guiding the next steps. The UNFCCC was thus designed in line with the notion that institutionalised international cooperation gathers momentum through a 'snowball' effect generating positive feedback and facilitating further steps. As noted in chapter 2, this process becomes more likely when initial institutional arrangements have a narrow scope, include lenient commitments and possess institutional feedback mechanisms that encourage dynamic development. These institutional qualities contained the seeds for progressing beyond the control of the fossil fuel industry.

Since 1989, the EU had worked for CO_2 stabilisation by means of firm reduction targets and timetables. The EU Commission aimed at taking a leading role in Rio by developing a climate 'package' containing a carbon/energy tax. While the EU-led initiative to set firm targets and timetables for CO_2 emissions went beyond the interests of the European fossil-fuel industry, the struggle within the EU prior to the Rio Conference provides a clear example of European industry influence. In general, business influence on EU environmental policy is substantial in all phases of the policy-making process (Grant et al., 2000). Depending on the decision-making procedures applied, EU industry has two main channels at its disposal: indirectly through domestic channels to national positions in the Council of Ministers, or directly through lobbying the Commission or Parliament. According to many observers, the carbon/energy tax proposed in 1991 was made subject to the most ferocious lobbying ever seen in Brussels (Skjærseth, 1994; see also Ikwue and Skea, 1994).

UNICE – the voice of European business at present representing 34 industry and employers' federations from 27 European countries – argued that the tax proposal ran completely counter to the need for concerted international action and would lead to difficulties for employment and competition (UNICE, 1991). While UNICE offered alternative climate strategies, EUROPIA played even tougher by expressing the 'strongest reservation concerning the creation of any new tax on oil products'.[2] Fierce industry lobbying led to the inclusion of tax conditionality, meaning that the tax would not be implemented if other OECD countries did not follow suit. This contributed to the killing of the tax and consequently to the crippling of EU leadership ambitions in Rio (Skjærseth, 1994; see also Newell, 2000). In the end, however, it was the member states that sank the tax proposal. Some countries argued that it did not go far enough. The Danes in particular said that it could not pass with the conditionality clause included. Other member states claimed that it was going too far, at too early a stage. The UK in particular opposed the use of fiscal mechanisms at the EU level as a matter of principle. Compared to the way the US fossil-fuel lobby killed the BTU tax proposed by the Clinton–Gore administration four years later (see chapter 5), the dispute about the EU carbon/energy tax also demonstrates the limits of industry influence within the EU.

Nevertheless, it was the US that determined the outcome of the UNFCCC: 'It was solely as a result of U.S. adamancy even in the face of complete isolation that the final text of the UN Framework Convention on Climate change (UNFCCC) contains only ambiguous language with regard to commitments' (Agrawala and Andresen, 1999: 461). Moreover, we have strong indications that industry played a key role in fuelling this adamancy prior to 1992. As discussed in chapter 5, the US pluralist political system provides ample room for interest groups to lobby for their interests in the development of national energy and climate policy. In the US, the domestic channel appears equally important for affecting international climate policies. The US fossil-fuel industry, in alliance with President George Bush, the Republican administration, and GHG sceptic chief of staff John Sununu influenced the mandate given to the US climate delegation (Newell, 2000). As early as 1989, the US delegation had been instructed to work for a weak framework convention on climate change through the United Nations Conference on Environment and Development (UNCED) process. According to Newell, 'the wording of the memo suggests that the negotiating space available to the US had been predetermined by a concern not to damage the interests of the fossil fuel industries' (Newell, 2000: 103).

Changes in corporate influence from Rio to Kyoto
Five years later, in December 1997, the Kyoto Protocol was adopted. The Protocol departed significantly from the interests of the fossil-fuel lobby and the oil industry in at least two ways: first, by requiring specific and mandatory reduction objectives within specific time frames ('targets and timetables'), and second, by exempting developing countries from any commitments. While Kyoto went far beyond the interests of the fossil-fuel lobby, it nevertheless fell short of the wishes of the green movement (Corell and Betsill, 2001). According to Grubb, 'the agreement was struck in the face of strong opposition from powerful industries, particularly in the United States, and with a set of flexibilities that were opposed by almost all environmental NGOs. It was very much an agreement struck by governments' (Grubb et al., 1999: 257).

Based on the UNFCCC, the Kyoto Protocol represented a visible sign of regime 'maturation' beyond the control of the fossil-fuel industry. With its commitment period from 2008 to 2012, the Protocol shaped expectations that could not be ignored by the fossil-fuel industry. In order to understand why ExxonMobil and Shell interpreted these expectations so differently, we have to understand how the driving forces underlying the match between industry interests and joint international climate commitments changed from 1992 to 1997. We will explore three propositions: (1) the international fossil-fuel lobby had dissolved and lost its influence prior to Kyoto; (2) the influence of the green movement had increased and thus outweighed the fossil-fuel lobby; and (3) changes in decision-making procedures and access served to reduce the power of the fossil-fuel lobby to influence decision-making.

The fossil-fuel lobby   The fossil fuel lobby is mainly represented through the GCC, which was established in 1989 and is characterised by many observers as the most powerful corporate lobby organisation in climate policy (Raustiala, 2001). The organisation represented a large share of US GDP. More than 50 big US and European companies based in the US participated, representing virtually every sector of US industry, including trade associations and the oil, coal, utility, chemicals and auto industries. ExxonMobil has been very active in the organisation.[3] The GCC spent a serious amount of money to convince policy-makers that proposals to limit CO_2 emissions are 'premature and are not justified by the state of scientific knowledge or the economic risks they create' (Levy, 1997:58). For example, the GCC sponsored a study concluding that the Kyoto Protocol would cost the US over 2.4 million jobs and reduce the GDP by US$300 billion annually (Carpenter, 2001).

After binding targets and timetables were prevented from entering the UNFCCC in 1992, the US fossil-fuel lobby continued to fight against them. US industry was also able to stop the BTU tax proposed by the Clinton–Gore administration in the Senate Finance Committee in 1993. At the first Conference of the Parties in 1995, the US agreed to the 'Berlin Mandate', declaring that non-binding commitments for developed countries were inadequate and that no new commitments would be imposed on developing countries. The latter had been approved by Vice-President Gore and was met with sharp criticism from industry lobbyists and the Republican Congress, subsequently forcing the administration to retract its position (Agrawala and Andresen, 1999).

In the period leading up to Kyoto, lobby activities intensified in the US. For example, the GCC sponsored a US$13 million television campaign saying that the price of petrol would increase by 50 cents per gallon if Kyoto timetables were implemented.[4] Before the final negotiations took place in Kyoto, the Senate delivered a clear message to the Clinton–Gore administration about its viewpoint: in a unanimous 95–0 vote, the Senate stated that it would not accept several of the conditions that are now codified in the Kyoto agreement. In July 1997 – five months before the Kyoto meeting – the Senate voted against any treaty that would exempt developing countries from legally binding commitments and imply higher energy costs, particularly on petrol (Byrd/Hagel resolution). For the Kyoto Protocol to be ratified by the US Senate, and thus become American law, it must receive 67 out of 100 senatorial votes. This non-partisan congressional resistance has been directly related to the powerful lobbyists representing the GCC in general, and the coal and oil industry in particular (Newell, 2000; Agrawala and Andresen, 2001). Since the Senate would have to ratify the Kyoto Protocol, the Byrd/Hagel resolution tied the hands of US negotiators in Kyoto and subsequently led the fossil-fuel lobby to believe that the Protocol would never be ratified in the US.[5]

In 2000, it was difficult to find any major US company that supported the Kyoto Protocol (Skodvin and Skjærseth, 2001). ExxonMobil, API and the GCC viewed the Protocol as 'dead on arrival', independent of the upcoming presidential election in 2000. The perception of the situation of these organisations before the 2000 election was as follows: the Democratic candidate, Vice-President Al Gore, would probably put all his political weight into persuading the Senate. However, Gore already had a low standing in the Senate and would in any case face considerable opposition. The Republican candidate, George W. Bush Jr, would have a higher standing in the Senate, but would not push for ratification.[6]

ExxonMobil has in fact been convinced that the US would never ratify the Protocol, and the company has firmly claimed that a 'Plan B' in case of ratification has never been needed.[7] The GCC also threatened lawsuits against legislation necessary to implement the Kyoto Protocol. The prospect of lengthy legal processes added to ExxonMobil's conviction that the Protocol would never become a reality in the US. The Protocol required the US to reduce its emissions by 7 per cent compared to 1990 levels within the period 2008–2012. Since 1990, US emissions have increased by over 14 per cent, and they are expected to increase by at least 1 per cent each year for the next decade. This means that the emissions will have to be reduced by more than 30 per cent during the agreement period. In principle, the Kyoto Protocol allows for emissions trading, but many doubted the political realism of the US becoming a major buyer of quotas in Russia. Russia's underground economy and poor overview of its own emissions would create serious problems for this kind of solution. Reductions of the order of 30 per cent by 2008–2012 seemed unlikely without large-scale investment in Russian quotas *and* powerful domestic climate policy instruments and measures. The US has no tradition of signing international treaties without a fair chance of compliance. If the US had ratified the Kyoto Protocol, the 2008–2012 targets would still be out of reach because of the need for dramatic domestic cutbacks combined with anticipated time-consuming legal action initiated by industry and environmental groups against the EPA. With the benefit of hindsight, the US withdrawal a few years later came as no surprise.

These domestic concerns did not, however, stop the international ambitions of the Clinton–Gore administration. The US position in the Kyoto negotiations was revealed shortly before the Kyoto summit in December 1997: President Clinton aimed to return GHG emissions to 1990 levels by 2008–2012, and participation by developing countries was set as a precondition for US agreement. The GCC and the API lobbied hard before Kyoto, and the API believed that it had succeeded in influencing the US position even though the stabilisation target went beyond API interests.[8] Meanwhile, the EU had agreed on an internal burden-sharing scheme and had proposed to cut emissions by 15 per cent from 1990 levels. Negotiations in Kyoto thus reached a deadlock. In this situation, Vice-President Gore made his famous 16-hour visit to Kyoto, where he called upon US negotiators to increase negotiation flexibility. The next day, the chief US negotiator signalled his willingness to negotiate emissions cuts that went beyond stabilisation (Agrawala and Andresen, 1999).

The EU position indicated that the European fossil-fuel industry had lost influence within the EU. European industry had been most concerned with the EU carbon/energy tax and consequences for European competitiveness. The shift in focus from taxes to burden-sharing, renwables and flexible mechanisms took care of the most controversial aspect of EU climate policy. UNICE, however, agreed with the US oil lobby that developing countries should participate in an effort to combat climate change. Moreover, UNICE has always stated that EU climate policy should be made conditional upon similar action taken by major trading partners (UNICE, 2000). The Kyoto Protocol, therefore, also failed to secure an international solution in line with the wishes of European industry, and it would be an exaggeration to say that the European fossil-fuel industry applauded the adoption of the agreement. European oil giants, however, did extend a cautious welcome to the Protocol, which may be seen as an early warning of the changes that were to come within the oil lobby.

Prior to the Kyoto meeting, the oil lobby was weakened by an emerging disintegration with BP's exit from the GCC in 1996, but the major changes in the strength of the coalition appeared after the adoption of the Kyoto Protocol: Shell followed BP in 1998. The exits of the European oil majors also seem to have set off a reaction in US companies: in late 1999, the Ford Motor Company withdrew from the GCC, followed by DaimlerChrysler, General Motors and Texaco. It is important to note, however, that even though the US companies could no longer accept the aggressive anti-climate stance of the GCC, they remained firmly opposed to the Kyoto Protocol, in contrast to the European oil companies (Skodvin and Skjærseth, 2001).

The GCC was 'deactivated' in 2002 – after 13 years in operation. According to the GCC itself, the group 'served its purpose by contributing to a new national approach to global warming', i.e. the US withdrawal from the Kyoto Protocol. Malicious tongues would say that the GCC has moved into the White House. However, another, and perhaps more important, cause is probably declining support within the business community, since the Protocol is still alive and future US participation cannot be excluded.[9]

Even though the main instrument for orchestrating the fossil-fuel industry opposition to GHG regulations – the GCC – did show signs of an emerging disintegration, there were no major changes in the alliance until *after* the adoption of the Kyoto Protocol. The EU position and the shift in focus away from taxes, however, shed light on the softening of the European industry position before Kyoto. The US fossil-fuel industry, on the other hand, remained unified and firmly opposed to binding targets and timetables. Moreover, the Byrd/Hagel resolution explains why

ExxonMobil was convinced that the Kyoto Protocol would never be ratified in the US. Since the US occupied a pivotal role in the climate regime, the county's fossil-fuel lobby also believed that a US withdrawal would put the Protocol out of action. Nevertheless, the change in match between fossil-fuel industry interests and joint commitments can only to a limited extend be ascribed to a weakening fossil-fuel lobby. On this basis, we would expect that other conditions had changed prior to 1997.

Counterbalancing forces: the green movement   Governments, target groups and ENGOs constitute the most important triangle of stakeholders present in the climate change process. ENGOs tend to represent a significant counterbalancing force to target groups, sometimes directly influencing their strategies as well as the policy of governments (see chapter 5). It is reasonable to assume that the more effective resistance to business strategies in this triangle of interests, the less industry influence can be expected. In contrast to chapter 5, we shall look specifically at whether *changes* in the influence of ENGOs over time shed light on changes in industry influence on joint climate commitments.

On the one hand, most analysts seem to agree that ENGOs have had little direct influence on the climate-change negotiations (Newell, 2000; Betsill and Corell, 2001). On the other hand, there seems to be an equally robust agreement that ENGOs have played an important indirect role by shaping the agenda, activating social demands and creating expectations. Carpenter (2001:320) even argues that ENGOs have always represented a formidable force in the climate-change negotiations. One major reason for these somewhat contradictory observations is that ENGOs' influence is extremely hard to measure. Also, they represent a diverse group that pursues a variety of goals and priorities. With these caveats in mind, we explore changes in ENGO influence in three decision-making arenas: the international negotiation process, and climate policy-making processes in the US and the EU.

In the *international negotiation process*, the final agreements do not reflect ENGO goals. Even though ENGOs constitute a diversified group, they were unified in their dissatisfaction particularly with the UNFCCC, but also with the Kyoto Protocol. Most ENGOs would agree that the Kyoto targets are too modest and that stronger review and compliance mechanisms are required. In

1997, with the slogan 'trading pollution is no solution', they also opposed emissions trading while they remained divided over the CDM, linked to forests and sinks. For example, many ENGOs strongly oppose the inclusion of any projects concerning biological sinks in the CDM, while other groups strongly support the inclusion of sinks offered by forest conservation and restoration.[10] As we shall see below, the ENGO position on emissions trading has changed significantly.

Lack of direct influence can hardly be traced back to lack of participation at the international level. The number of all types of NGOs (including business and industry groups and ENGOs) accredited as observers more than doubled between 1992 and 2000 (Carpenter, 2001). More than 40 ENGOs sent representatives to at least two of the nine sessions of the Ad Hoc Group on the Berlin Mandate (AGBM) and on average more than 100 ENGO representatives were at each session (Corell and Betsill, 2001). However, ENGOs were outnumbered by industry: at AGBM-6, there were about 35 business organisations with some 150 representatives, compared to 25 ENGOs with about 90 representatives (Raustiala, 2001). At these sessions, ENGOs tried to coordinate their strategies under the Climate Action Network (CAN). Established in 1989, CAN is an umbrella organisation comprising most international ENGOs that are active on climate-change issues in central and eastern Europe, Latin America, South East Asia, South Asia, Africa, the US and Europe. CAN Europe (Climate Network Europe, or CNE) alone enjoys the membership of 85 European ENGOs. At the AGBM and Conference of the Parties (COP) meetings, CAN members held seminars, interacted with the media and lobbied delegates.[11]

Despite increased participation, the influence of ENGOs appears to have declined over time. First, in the late 1980s, ENGOs were very influential in setting the agenda of climate change as an issue that had to be addressed. At that time, most industry organisations had limited knowledge of both the problem and the consequences of problem solving. ENGOs actually dominated the ranks of NGOs in the emerging climate regime, but have faced increased opposition from business and industry groups over time. Second, ENGOs have become increasingly diversified (Carpenter, 2001). An increasing inability to 'speak with one voice' has constrained their power to influence negotiation processes. Third, while ENGOs were permitted access to most negotiating arenas, they were excluded from informal meetings where the real negotiations took place. From the INC meetings during the initial phases of the process to the Kyoto negotiations, ENGOs were excluded from closed-door sessions (Newell, 2000). The most effective channel for ENGOs as well as for the fossil-fuel lobby, however, is the domestic.

In the *US*, ENGOs were instrumental in putting climate change on the political agenda in the late 1980s. In 1988, the first element of a climate regime emerged at an international conference in Toronto calling for a 20 per cent cut in CO_2 emissions by 2005. US environmental groups participated at this conference and used the international process to market the threat of climate change to domestic actors (Agrawala and Andresen, 1999). This happened before the US fossil-fuel industry had mobilised and formed the GCC. Moreover, it happened concurrently with the heat waves and drought that hit North America in the summer of 1988. According to Agrawala and Andresen (1999), however, the influence of US ENGOs was significantly reduced as the international process leading up to the UNFCCC became more institutionalised and the fossil-fuel lobby grew in power. These researchers further argue that US ENGOs today function primarily as disseminators of information on international negotiations and assessments. This is supported by the fact that Greenpeace-US resigned from its efforts to lobby for US ratification even before Bush Jr was elected president.

The strategies chosen by the environmental groups may also have undermined their influence when the fossil-fuel industry mobilised: US ENGOs adopted a climate-change strategy based on cooperation with industry (Eikeland, 1993). Newell (2000) argues that the low influence of ENGOs in US climate policy is related more to strong opponents than to lack of access to decision-making processes. Indications of real influence with possible consequences for joint commitments are mainly related to the creation of public expectations that have contributed to participation at a high level. There seems to be a quite robust pattern indicating that the green movement in North America has experienced a decline in influence on US climate policy.

In the *EU*, few green organisations are permanently represented in Brussels. However, ENGOs are represented by eight large umbrella groups such as the CNE, Greenpeace and the European Environment Bureau (EEB). There are fewer people representing environmental interests in Brussels as a whole than there are representing business interests in the UNICE secretariat alone. While noting that industry groups are generally in a stronger position within the EU policy-making process, Grant et al. (2000) also argue that the supranational EU system provides ENGOs with some specific agenda-setting opportunities. Of particular importance is the fact that the environment directorate in the Commission needs ENGOs to do its job. The small size of the Commission makes it dependent on ENGO sources for advice. The Commission contributes financial support to ENGOs, and the EEB was founded by the Commission as a counterbalancing force to industry groups. The symbiotic relationship between the environment directorate and ENGOs has also been reflected in the strategies of the European ENGOs: they seek to strengthen and expand proposed targets and measures, and they support legislative proposals from the Commission against the Council and business interests (Skjærseth, 1994).

The problem for the ENGOs, however, is that industry tends to make alliances with more powerful parts of the Commission. The dispute about the EU carbon/energy tax is quite illustrative. The driving force within the Commission was an alliance between the directorates for environment and energy. This was supported by ENGOs such as Greenpeace and Friends of the Earth (FoE). For example, FoE presented a position paper defending the tax proposal against industry.[12] On the other side, the industry allied with the most 'business-friendly' and powerful directorates responsible for economics and the internal market. This alliance paved the way for the principle of conditionality (Skjærseth, 1994).

Significant changes have occurred since the European ENGOs were defeated over the carbon/energy tax proposal. While the US concept of emissions trading was met with a wall of ENGO protests in Kyoto, emissions trading has now been accepted by European green organisations. CNE, for instance, 'regards the Directive [the EU directive on emissions trading] as a potentially useful proposal' and argues that it 'offers the potential for significant emissions reductions within the European Union, [although it] contains a number of actual and potential weaknesses'.[13] CNE argues that the EU scheme should be kept separate from Kyoto trading in order to exclude 'hot air' and sinks. A more fundamental change is perhaps the emerging agreement and cooperation between European business, industry and ENGOs, demonstrated particularly in relation to the development of a scheme for emissions trading within the EU. In essence, while EU climate-change policy in the early 1990s pitted all stakeholders against each other over an EU proposal (the carbon/energy tax), a US proposal (emissions trading) unified the same stakeholders in 2001.

While ENGOs certainly have been present in the decision-making process at the international level as well as in the US and EU, there are no indications that their influence has increased over time. On the contrary, to the extent that there has been a change in ENGO influence, it is in the direction of a decrease rather than an increase. Thus, the fossil-fuel industry has not lost control over the international negotiations due to increased ENGO influence. Change in the influence of the environmental movement can hardly explain the changes in the influence of the fossil-fuel lobby from the UNFCCC to the Kyoto Protocol.

Access and decision-making procedures   In general, non-state actors have had good access to the international negotiation process, although there are indications that the scope for influencing the process appears to have been higher in the run-up to the UNFCCC than to the Kyoto meeting (Corell and Betsill, 2001). As noted above, however, non-state actors have regularly been excluded from participation at informal meetings. Particularly during the final negotiations of the Kyoto Protocol, negotiation meetings were increasingly closed for non-state delegates. Thus, while there are some indications that there were changes in non-state actors' access to decision-making processes at the international level between 1992 and 1997, these minor shifts cannot fully account for the significant alteration in industry influence during this period. Access to the international meetings is not perceived as critical by the fossil-fuel lobby. The API estimates the relative importance of domestic versus international channels as 70 to 30.[14]

However, the US oil industry experienced a temporary change in the US administration that implied less access for the industry to relevant decision-making processes at home. In general, there was little cooperation and consultation between the Clinton–Gore administration and fossil-fuel industry groups before the Kyoto negotiations. In particular, the US oil industry was excluded. The API's perception is that the Clinton–Gore administration showed no interest in cooperating with 'Big Oil'.[15] API and other industry interests met with Gore in Kyoto but were unable to change his view.[16] Vice-President Gore's call for increased negotiation flexibility in Kyoto thus prevented the fossil-fuel industry, in alliance with Congress, from inflicting its interests upon the rest of the world.

US flexibility generated concessions from other countries leading to the consensus agreement on the Kyoto Protocol and the US commitment to reduce GHG emissions by 7 per cent from 1990 levels by 2008–2012 (Agrawala and Andresen, 1999). The US also relented on the issue of participation by developing countries, but won a number of other concessions, particularly from the EU.[17] Ironically, the Kyoto Protocol, with its emphasis on sinks and flexibility mechanisms, is essentially a US construction designed to sugar the pill for US industry.[18] In the event, however, the US commitment to Kyoto represented only a temporary loss of US industry influence. The industry maintained a strong influence in Congress through the Byrd/Hagel resolution.

For the Kyoto Protocol to enter into force it must be ratified by 55 countries representing at least 55 per cent of global CO_2 emissions in 1990. The US is responsible for about one quarter of the global emissions of GHGs and is thus a key actor in international climate politics. The Protocol can enter into force even without an American ratification, but the central role of the US implied that a Kyoto Protocol without it was not seen as very realistic. Thus, US multinationals with significant activities in Europe still had a fair chance of exercising influence beyond their home-base country.

In addition to access, *decision rules* represent another important determinant for industry influence. The climate regime itself has been based on consensus. All decisions under the Convention need to be made by consensus since it has been impossible to adopt rules of procedure. This implies that the US fossil-fuel industry nearly blocked the adoption of the Kyoto Protocol through the Byrd/Hagel resolution.

Within the EU, an institutional development with significant implications for European businesses' influence over EU climate policy has taken place. A development towards an increased use of qualified majority voting in the Council of Ministers has by itself affected the organisation of the EU lobby. Since the Single European Act of 1987, decisions related to the internal market have been based on Article 100a, which requires a qualified majority. This implies that member states can be outvoted in Council, and consequently that the domestic channel is no longer sufficient for lobby organisations. As a consequence, the number of lobby organisations representing business and industry in Brussels rose dramatically, from about 600 in 1986 to almost 3,000 in 1990 (Skjærseth, 1994). Thus, organised lobbying of the EU Commission is a more recent phenomenon in the EU than in the US.

Complex decision-making procedures have evolved in the EU whereby the European Parliament has been given more power and there is greater use of majority voting within the Council of Ministers. The European Parliament is widely perceived as the 'greenest' of the EU institutions. The 1992 Treaty of Maastricht extended the use of majority voting so that it became the rule on environmental matters. Fiscal measures, however, were excepted from this procedure. The proposed carbon/energy tax would thus require consensus, meaning that European businesses would have to persuade only one state to block the proposal. The 1997 Amsterdam Treaty broadened the application of the co-decision procedure to cover directives adopted on the basis of the 'environmental' paragraph 130s, hence increasing the decision-making role of the Parliament further. These institutional changes and the shift in focus away from taxes prior to Kyoto thus narrowed the scope available for European business organisations to influence EU climate policy, with a softened industry opposition as a result. Relevant EU directives since Kyoto have been proposed on the basis of qualified majority. For example, the EU directive on a framework for GHG emissions trading is proposed according to Article 175 (1), which requires a qualified majority (see below).

Summary: corporate influence
The US fossil-fuel industry, with ExxonMobil in the lead, did not lose influence on joint climate commitments due to a weakening of the fossil-fuel lobby or stronger counterbalancing forces in the

green movement. If that had been the case, a company like ExxonMobil would have had to reconsider its reactive climate strategy. While the fossil-fuel lobby does show signs of an emerging weakening during the period immediately preceding the adoption of the Kyoto Protocol, the major changes in the strength of the fossil-fuel coalition occurred after the adoption of the Kyoto agreement. Moreover, even the US multinationals that left the GCC continued to oppose the Kyoto Protocol. Similarly, while the environmental movement certainly has been very much present in the development of climate policy both at the international level and in decision-making processes in the US and the EU, its influence has decreased rather than increased during the period from 1992 to 1997.

Within the EU, a permanent change in the decision-making procedures took place with the increased use of decision rules based on a qualified majority and more involvement by the Parliament in decision-making. This change implied that it was no longer sufficient for the industry to persuade only one member state to bar unwanted policies. To influence decision-making, the reactive industry had to convince a majority of the member states, which is a more difficult task. In addition, the change in EU climate policy from taxes to burden-sharing, renewables and flexible mechanisms reduced European companies' resistance to GHG measures. Also, BP's exit from the GCC before the adoption of the Kyoto Protocol weakened the oil lobby, although the major change in the cohesion of the oil lobby took place after the adoption of the Protocol. Thus, the climate policy of the EU acquired its own dynamic largely beyond the control of big European multinationals like Shell.

In the US, the fossil-fuel industry was excluded from the decision-making process of the Clinton–Gore administration in the period immediately preceding the Kyoto negotiations. In contrast to the changes that occurred in the decision-making procedures in the EU, however, this loss of influence was only temporary. The Byrd/Hagel resolution implied that the industry still exerted a strong influence in Congress, which had the final word concerning US ratification of the Kyoto Protocol. Here, we also find an important reason why ExxonMobil did not modify its climate strategy as a consequence of the US signing of the Kyoto Protocol: ExxonMobil and the rest of the US fossil-fuel industry were convinced that the Kyoto Protocol would never be ratified in the US, and as a consequence probably not ratified at all. A Kyoto Protocol in force without the United States could not be excluded, but would represent a dramatic weakening of the climate regime. In essence, the US fossil-fuel industry maintained strong influence over the climate regime in either case.

The Kyoto Protocol departed significantly from the interests of the fossil-fuel industry. This development was apparently caused by an evolving regime set in motion by the UNFCCC. The US fossil-fuel industry in particular made intense efforts to block this process through domestic channels, but did not succeed. However, this industry knew it could block US ratification at a later stage. Significant changes in the strategies of the industry occurred after Kyoto and mainly in Europe.

From Kyoto to Marrakech: regime impact on corporate strategy choice

As it turned out, the adoption of the Kyoto Protocol became the precursor to an emerging divide between Europe and the US – both in policy measures and in fossil-fuel industry strategy.

In this section, we analyse the extent to which the changes in corporate strategies that took place after the adoption of the Kyoto Protocol can be traced back to changes in the international climate regime. We home in on the emerging division between the international climate regime and the EU and explore the causal relationship between regime features and corporate strategies. The discussion is organised around the three causal pathways through which international regimes can affect corporate strategy that were discussed in chapter 2: knowledge, pressure and opportunity.

Knowledge

The scientific uncertainty argument has repeatedly been used within the corporate fossil-fuel lobby. ExxonMobil builds its reactive climate strategy on the assumption that the international climate regime is based on 'bad' science, while Shell accepts the IPCC's definition of the problem. The assumptions put forth in chapter 2 were that differences along this dimension might result from differences in corporate access to the IPCC process.

Alternatively, there might be differences in the companies' receptiveness to the information provided.

The main institution for the generation of a common understanding of the nature of the climate problem is the IPCC, established under the auspices of the United Nations Environment Programme (UNEP) and the World Meteorological Organisation (WMO) in 1988. The purpose of the body is to give all concerned parties – governmental as well as non-governmental – equal access to the most up-to-date, state-of-the-art scientific knowledge on climate change. The IPCC thus has an independent institutional platform, separate from the political bodies of the climate regime, which are organised directly under the UN General Assembly (see Skodvin, 2000a, 2000b). With its intergovernmental status, however, the IPCC is nevertheless a hybrid organisation operating at the interface of science and politics. This status implies that the body is open to participation by all UN member states.

With its independent institutional platform, the IPCC is neither under the direct control of the political bodies of the climate regime, nor has any channels for direct input to the bodies where political deliberations take place. One of its main functions, however, is to develop – through negotiations between scientists and government officials – a common understanding of the nature and magnitude of the climate problem and possible solutions (Skodvin, 2000a). This implies that the IPCC can affect climate policy in at least two important respects. Its work can have an impact at the national level, to the extent that national decision-makers accept IPCC conclusions as valid and act upon them in their design of national climate policies. The IPCC can also have an impact at the international level, to the extent that it succeeds in developing a problem definition, agreed upon by parties representing conflicting interests, which thus can constitute a common framework of understanding upon which international climate policies can be based. Thus, even though the institutional set-up of the IPCC implies that it does not have direct 'access' to policy-making, its work and the associated process constitute an important part of the framework conditions within which the oil industry develops its climate strategy (for a more detailed analysis of the IPCC see, for instance, Agrawala, 1998a, 1998b; Skodvin, 2000a, 2000b).

The IPCC has issued three assessment reports – in 1990, 1995 and 2001 – in addition to a number of technical papers and special reports (issue-specific assessments). Since the first assessment report (1990), the IPCC has drawn strong conclusions about the (causal) relationship between human-induced emissions of GHGs and a trend of increasing global mean temperatures, although it is fair to say that this attribution of climate change to human sources has been increasingly scientifically substantiated during the course of the process. Thus, rather than progressing by sudden scientific breakthroughs, the IPCC process has taken the form of a gradual development of the knowledge base towards increasingly strong conclusions about the causal relationship between anthropogenic GHG emissions and climate change.

Climate science, however, is also associated with major points of uncertainty that may not be resolved for quite some time. Accounts of the confidence scientists have in climate model simulations and projections indicate that while scientists consider it virtually certain that the anthropogenic increases in atmospheric GHG concentrations will affect the climate for many centuries, they are uncertain about the precise nature of the response of the climate system, the rate of change, and the attribution of the observed warming trend to anthropogenic increases in the atmospheric concentrations of GHGs (see, for instance, Mahlman, 1997).

As noted, while the US fossil-fuel industry, particularly as represented by ExxonMobil and the GCC, did not accept the problem definition provided by the IPCC, the European-based industry did. The *responses* of these industries to the IPCC knowledge base vary accordingly. The US-based industry, with the GCC in the lead, tried to influence the development of the knowledge base itself and discredit the scientific authority and legitimacy of the scientists who provided it. It could be argued that 'sceptics' (herein also the oil lobby) have employed increasingly aggressive methods in this endeavour during the course of the IPCC process. The most prominent example of this is the debate that followed in the aftermath of the IPCC's endorsement of the second assessment report. The IPCC's re-editing of this report led to accusations from the GCC of 'scientific cleansing'.[19] Moreover, in an editorial-page piece in the *Wall Street Journal*, Ben Santer, who was convening lead author of the most controversial 'attribution-chapter' of the second assessment report (chapter 8), was held personally responsible for the 'cleansing'.[20] Santer interpreted the incident as 'a skilful campaign to discredit the IPCC, me and my reputation as a scientist'.[21]

These lobbying tactics represented a more aggressive strategy by the coalition. While it certainly served to draw the attention of international media, it also backfired in the sense that it served to alienate some of the GCC's own members. In 1996, *Nature* reported that 'several GCC member companies are understood to have been uneasy about the organization's aggressive tactics'.[22] For instance, the aggressive lobbying tactics alienated the European-based industry, which acknowledged to a much larger extent the problem definition provided by the IPCC, and accepted the climate problem as a concern that needed to be addressed and taken into account in its operations.

There are at least two possible explanations for these variations. First, they may have been caused by corresponding variations in the companies' access to – and hence influence over – the IPCC process and its conclusions. Second, they may have been caused by company-specific differences in the companies' receptiveness to the knowledge and information provided by the IPCC.

The IPCC is organised in three main, cross-cutting decision-making levels: (1) the 'scientific core', which is responsible for producing the actual scientific assessments of each of the three working groups (WGs); (2) the WG Plenaries, at which the IPCC formally endorses the bulk assessment reports; and (3) the highest decision-making level, which is the full Panel Plenary, whose main tasks are organisational and administrative in nature (except for the formal endorsement of a Synthesis Report, in which the main conclusions of the reports from all the three WGs are synthesised).

At WG and Panel Plenary levels, the IPCC is characterised by an unprecedented openness. While it is only national delegates (and scientists, in the endorsement of summaries at WG level) that have formal voting power in the IPCC, all NGOs and intergovernmental organisations (IGOs) accredited within the UN system can participate at IPCC meetings. At these levels, these meetings have never been closed to participants from these categories, but speech restrictions have been enforced at meetings with a tight time schedule. The oil industry has participated actively in the IPCC process through a variety of organisations, but the GCC has had a leading role in the orchestration of the lobby's opposition. Variations in access as an explanatory factor for variations in industry responses to the knowledge base provided by the IPCC do not, therefore, have much support in this case. Companies seem to have had equal opportunities to participate in the process.

Company-specific variations in receptiveness to the knowledge provided by the IPCC, however, seem to gain more support in this case as an explanation of variations in industry responses. One indication is the difference between ExxonMobil and Shell in the extent to which the companies themselves conduct research on the areas covered by the IPCC (see also chapter 4). This difference is emphasised by officials from both companies.[23] Exxon has had internal expertise since the early 1980s, and made further investments in R&D activities in this area during the 1990s.[24] In contrast, Shell has gradually decreased its R&D activities over the last couple of decades and has had its own climate expertise only since the very late 1990s.[25]

The distinction between ExxonMobil and Shell in their internal climate expertise is also reflected in differences in the nature of their participation in the IPCC process. While both Shell and Exxon representatives have participated as observers at IPCC meetings and as reviewers in IPCC assessment reports, only Exxon scientists have participated as cited contributing authors listed in the references of the reports. This indicates that there is a clear difference between the companies in their dependency on IPCC conclusions. While Shell has limited internal climate expertise upon which to base counter-claims, ExxonMobil has a relatively long tradition of climate science and can thus present competing knowledge claims based on their own research (see also Kolk and Levy, 2001).

The success of the GCC strategy in terms of dissuading the new Bush administration from taking on Kyoto commitments should be noted, however. This is reflected in Bush's rhetoric in relation to the US withdrawal. When he rejected the Kyoto agreement, he called it 'fatally flawed',[26] and also reportedly 'questioned what he called the "incomplete state of scientific knowledge behind it"'.[27] In spring 2002, moreover, events suggest that ExxonMobil successfully lobbied the White House to withdraw US support of Robert Watson's continued chairmanship of the IPCC.[28]

The IPCC conclusions thus brought about different perceptions within the fossil-fuel industry on the nature and urgency of the climate problem – with a main distinction running between European and US-based companies. It is important to note that there were no breakthroughs in climate science immediately preceding the strategy changes that could have induced or persuaded the industry to change its strategy.

Regime pressure

In spite of strong opposition from the fossil-fuel lobby, parties to the UNFCCC succeeded in adopting the Kyoto Protocol – a regulatory framework for international GHG regulation that certainly represented a reinforcement of the more lenient UNFCCC. Moreover, with the adoption of the Buenos Aires Plan of Action at COP-4 in 1998, parties succeeded in institutionalising the continuation of the cooperation even before the details of the agreement had been worked out. Thus, with the Kyoto Protocol and the decisions following in its aftermath, parties succeeded in racking up the international climate regime a notch, as well as in providing institutional feedback mechanisms around which actor expectations converged.

The dynamic these achievements set off, however, took on a completely different form in the US and in Europe. In the US, the agreement backfired and ultimately brought US climate policies back to square one (the UNFCCC level of ambition). In Europe, the adoption of the Kyoto Protocol set off a process through which the stringency, specificity and level of ambition of European GHG regulations were gradually further increased.

The response to the Kyoto Protocol by the US industry is – *to some extent* – dual. There are indications that parts of the US industry did take initial steps towards a more proactive stance. One example of this is the foundation of the Pew Center on Global Climate Change – a US thinktank – in 1998. The objective of the centre is to educate policy-makers as well as the general public about the causes and possible consequences of climate change and to encourage the domestic and international communities to reduce emissions of GHGs. In 2000, the Pew Center comprised 21 major US companies. Today, the number has risen to nearly 40, most of which are included in the Fortune 500. The Pew Center now comprises previous GCC members such as Shell and BP. The problem, however, was that the entire US industry – including Pew members – was opposed to the Kyoto agreement. Thus, not even Pew had an explicit position on the Kyoto Protocol, at least not initially.

Moreover, for the other – and as it turned out, the major – part of the US industry, the Kyoto agreement did not induce small steps towards a more proactive stance at all. On the contrary, the industry almost felt betrayed by the US government's consent to the agreement,[29] and its adoption set off a reinforced mobilisation against it. This strategy was crowned with success when the newly elected Bush administration announced its withdrawal from the Kyoto Protocol in 2001. After this, the international climate regime entered the doldrums.

In Europe, on the other hand, the adoption of the Kyoto Protocol – and the EU's commitment to reduce its emissions by 8 per cent from 1990 levels by 2008–2012 – set off a completely different dynamic. For the EU, the Protocol represented the 'green light' it needed to proceed with its development of climate policy. The direction in which the EU was heading on this issue had been evident for quite some time: towards a fully fledged regulatory framework for GHG emissions abatement. As the EU is a supranational entity, its decisions, in the form of legally binding directives and regulations, have a higher authoritative force than decisions in regular intergovernmental cooperation. Also, EU decisions tend to be linked to stronger verification and compliance systems.

The development of the EU climate regime was gradually delinked from progress (or the lack thereof) in the international arena. The EU worked along two increasingly independent pathways: one that constituted efforts to save and strengthen the international climate regime in order to get other countries at least to do *something* on climate change (EU as actor); and another that constituted EU implementation of its Kyoto Protocol commitments as if they had already entered into force (EU as regime). As the latter process progressed, it gained its own momentum and gradually became independent – at least temporarily – of the success or failure of efforts along the other pathway.

In June 1998, the EU adopted the renegotiated burden-sharing agreement. Thus, both the EU and the member states will have legally binding targets and share the responsibility for meeting them. In March 2002, EU member states agreed to be legally bound by the Kyoto Protocol, and the EU and its member states ratified the agreement in May 2002.

In 2000, the EU established the European Climate Change Programme (ECCP) to 'drive forward EU efforts to meet the targets set by the Kyoto Protocol' (ECCP, 2001: 3). With its multi-stakeholder approach, the ECCP serves both as an instrument for ensuring progress in the implementation of EU climate policies, and as a vehicle for participation by industry in the process. Also, the EU has proposed an emissions trading directive as the core element of the ECCP. In December 2002, EU environmental ministers agreed to create the world's first international GHG emissions trading system, subject to final approval by the European Parliament. The EU has had a monitoring mechanism for its GHG emissions since 1993.

The European-based industry responded to these changes in political framework conditions with an almost unequivocal support of the Kyoto Protocol. In parallel with the development in EU climate policies, a *reactive* stance on the issue became a less viable option for industries most affected by the EU regulatory framework. At the very least, such a strategy was increasingly associated with a significant risk. Moreover, the risk associated with a *proactive* strategy was reduced because new markets and opportunities had been developed as part and parcel of the European climate regime. At the same time, this shift in the strategies adopted by the European oil industry further facilitated the EU's drive towards a more ambitious climate policy. The establishment of in-house emissions trading schemes within BP in 1998 and Shell in 2000 acted as key drivers in the EU policy debates (Christiansen and Wettestad, forthcoming). Thus, the relationship between the European industry and EU political authorities had shifted from contentiousness during the fight over the carbon/energy tax in the beginning of the 1990s to a more harmonious situation a decade later, even though there are still different views on specific parts of the ECCP. Currently, the European industry, particularly the oil industry, to a much larger extent represents a partner in the development of EU climate policies – a situation particularly evident in the process leading up to the directive on emissions trading.

This dynamic, however, should also be seen in relation to the development of climate policies at the national level. Within the EU, one of the most proactive countries on the climate issue is the Netherlands. As discussed in chapter 5, Shell sees the Netherlands as a 'test country' for what the industry can expect from a viable climate policy.[30] In 1989, the Dutch government announced its decision to stabilise CO_2 emissions at 1989/1990 levels by 1995 at the latest. In 1990, a revised plan called for a 3–5 per cent reduction from average 1989/1990 levels by 2000. The Netherlands was among the first major industrialised countries to ratify the Kyoto agreement.[31] Simultaneously, policy instruments have been stepped up. Shell, therefore, has operated within a political context that has sent strong signals of future GHG regulations from two levels: the national level in Dutch climate policies and the regional level in EU climate policies.

As framework conditions for industry, the international and EU climate regimes thus have markedly different consequences. In the international climate regime, ambiguity and uncertainty have prevailed – and even increased with the US exit – throughout the regime-building process. In contrast, EU climate policy-making has maintained and even increased its momentum during the last decade, thus sending a strong signal to industry of an emerging regulatory framework on this issue. After a faltering start during the early 1990s, the EU succeeded in providing a leadership role in the climate process during the latter half of the decade. The EU had to admit defeat in persuading the US not to withdraw from the cooperation and mobilised support to 'go it alone'.[32]

Regime opportunities

The EU climate regime served to create new market opportunities particularly in two areas: the development of a market for renewable energy sources and the development of an internal EU scheme for CO_2 emissions trading.

Renewable energy sources   In implementing its Kyoto Protocol commitments, the EU has identified a series of energy actions, including a prominent role for renewable energy sources. The development of renewable energy has for some time been a central aim of Community energy policy. As early as 1986, the Council listed the promotion of renewable energy among its energy objectives, and with the ALTERNER programme (1992) the Council adopted a specific financial instrument for renewables promotion (COM (97) 599 final).

The first step towards a strategy for renewable energy was the Commission's adoption of a Green Paper in November 1996. In 1997, the Commission adopted a strategy and action plan directed towards the goal of having renewable energy sources cover 12 per cent of the EU's gross inland energy consumption by 2010 (COM (97) 599 final). In its decision of 28 February 2000,[33] the European Parliament decided to extend the ALTERNER programme in the establishment of ALTERNER II as a 'specific programme for promotion of renewable energy sources'. The financial framework for the implementation of the programme for the period 1998–2002 is set at EUR 77 million. Both UNICE and EUROPIA support the EU's investments in renewable energy.

The long history of the renewables policy within the European Community has been marked by financial setbacks (see Wettestad, 2001). Nevertheless, ALTERNER and particularly the establishment of a financial mechanism in 1992 are an indication of a political effort to develop a market for renewables in Europe (see Ikwue and Skea, 1994). This may have been the political signal the European-based oil industry needed to initiate its own investments in renewable energy. There is thus a noteworthy coincidence between events taking place on the policy side during this period and events within the European-based oil industry. The change in BP's position on the climate issue, for instance, is remarkably parallel with the Green Paper issued by the Commission in 1996, where the 12 per cent target on renewable energy sources was tentatively suggested. Similarly, the following White Paper, where the 12 per cent target was adopted, parallels Shell's establishment of a fifth core business area – Shell International Renewables – in 1997.

Again, however, the relationship between the EU and Shell should be seen in the light of the development of renewables in the Netherlands (see also chapter 5). In 1995, the Netherlands adopted a target of renewable energy sources attaining a 10 per cent share of total energy consumption by 2020. Over the past few years, policies to promote renewables have been strengthened by the introduction of a wide range of fiscal arrangements, such as fiscal tax credits. Renewable energy is also promoted through its inclusion in the long-term agreements (VROM, 1999). Thus, Shell was subjected to 'double exposure': both the Dutch and the EU renewables targets directly influenced its decision to establish its fifth core business area in 1997 (Skjærseth and Skodvin, 2001).

These developments may not, however, have been enough to push the industry from a reactive to a proactive position. They do indicate, though, that there were movements in the industry's position even before the adoption of the Kyoto Protocol. The actual *adoption* of the Protocol may have been the crucial factor that tipped the balance, pulling the industry towards a more proactive strategy.

The European emissions trading scheme   Another central component of EU implementation of its Kyoto Protocol commitments is the internal EU scheme for emissions trading. In June 1998, the Commission stated that 'the Community would set up its own internal trading regime by 2005', three years prior to the operation of an international trading regime (cited in COM (2000) 87 final: 10). As noted, the Council of Ministers has agreed on the proposal for an EU directive on a regulatory framework for GHG emissions allowances within the Community (COM (2001) 581 final).

The emissions trading scheme was prepared in accordance with the ECCP's multi-stakeholder approach. Thus in September 2001, a stakeholders' consultation meeting was convened that included all major business and industry organisations – such as UNICE, the European Roundtable of Industrialists and EUROPIA – as well as representatives of the environmental movement. The Summary Record of the meeting reveals a broad-based consensus on the main issues, including an overwhelming majority in favour of going ahead with emissions trading sooner rather than later, and a general agreement on a mandatory scheme from 2008.[34]

The number of installations covered by the EU emissions trading scheme proposal are limited to some 4,000–5,000 covering about 46 per cent of (total) CO_2 emissions of the 15 EU member countries in 2010. Refineries are included in the core activities, thus making the oil industry an important target group. The change in the position of EUROPIA illustrates the significant shifts that took place within the European oil industry after the Kyoto Protocol. EUROPIA is composed of 21 oil companies and represents the downstream activities (including refineries) of the oil industry within the EU institutions. In the early 1990s, EUROPIA was, as discussed earlier, one of the most aggressive opponents of EU climate policy, particularly the carbon/energy tax. Today, EUROPIA welcomes the EU emissions trading directive as a learning exercise towards an international system under the Kyoto Protocol. In general, EUROPIA supports UNICE's position on climate change (EUROPIA, 1999). And UNICE supports the proposed EU directive as a 'learning-by-doing' process in the period before 2008. In a detailed position paper on the EU proposal, UNICE has no comments on mandatory monitoring, reporting and verification requirements. With respect to enforcement, UNICE supports financial penalties, but proposes to change the wording in the directive to avoid unnecessary uncertainty (UNICE, 2002).

There are at least two closely related reasons underlying the change in the position of EUROPIA: changes in the climate policy of the EU and its member states, and changes in the strategies of Shell and BP particularly.[35] With respect to the former, the prospects of a stringent internal EU scheme for emissions trading from 2005 sent a strong signal to industry that a new market for CO_2 was emerging and that industry should prepare itself. And with respect to changes in Shell and BP, the parallelism in the Green Paper issued by the Commission on the design of an emissions trading system and Shell's launching of its own internal emissions trading system (STEPS) – both events taking place in 2000 – is interesting. These policy measures serve to generate new market opportunities and may decrease costs of emissions reductions. It is illustrative, for instance, that BP's three-year-old emissions trading system yielded £650 million in extra value for the company.[36]

EUROPIA also includes US companies operating in Europe, such as Texaco-Chevron, Conoco, Phillips and ExxonMobil. In fact, ExxonMobil has a leading role within EUROPIA even on environmental matters.[37] This shows that EUROPIA's position is determined by the corporate leaders in this field and not the lowest and most powerful denominator.

Conclusion

In this chapter, we have shifted our focus from the domestic level of the DP model to the international level of the IR model. The main contribution of the latter is that it allows for a stronger focus on the *dynamic relationship* between corporate strategy and international institutional development and thus serves to improve our understanding of *changes* in corporate climate strategy choice. In particular, two questions remained unanswered from our analysis in the preceding chapters. What caused Shell's turnabout from a reactive to a proactive strategy in 1997/1998? And why did ExxonMobil not modify its reactive strategy in response to its apparent lack of influence resulting in the US signing of the Kyoto Protocol in 1997?

Recalling our discussion in chapter 2, we assumed that the extent of influence corporate actors have on international regimes constitutes an important determinant for corporate strategy choice. Corporate influence is at least linked to three factors: (1) the cohesion of the industry group and its ability to act in a unified manner; (2) the strength of counterbalancing forces, in particular ENGOs; and (3) the decision-making procedures applied and the group's access to central decision-making processes. If a reactive industry exercises a strong influence and largely controls the development of the regime, a persistent reactive strategy can be expected. In contrast, if the industry has a weak influence and the regime 'matures', we can expect a change towards a proactive strategy.

Our analysis shows that neither the emerging dissolution of the oil lobby nor a change in the influence of the green movement can provide satisfactory explanations for the loss of industry influence on international climate commitments from 1992 to 1997. Rather, the decreased influence of the fossil-fuel industry preceding the adoption of the Kyoto Protocol was caused by a number of factors: a change in industry access at the governmental level in the US; a change in EU climate policy from taxes to burden-sharing, renewables and flexible mechanisms; and a change in decision rules within the EU. The adoption of the Kyoto Protocol, moreover, became the precursor to the emerging US–EU divide in both climate policies and industry strategies. The European-based industry shifted from a reactive to a more proactive strategy in response to EU regulatory pressure and new market opportunities, on the basis of what was perceived as a persuasive knowledge base.

In the run-up to the UNFCCC, the industry exerted a strong influence on the development of the climate regime, indicated by the match between industry interests and the joint commitments of the regime: the UNFCCC reflected the interests of the fossil-fuel industry at large, particularly in the sense that no legally binding reduction targets and timetables were included. During this phase, the fossil-fuel lobby, constituted by both European- and US-based companies, represented a cohesive and unified coalition. This exerted a relatively strong influence on the climate regime, but primarily at the national and regional levels. As we have seen, the US industry influenced the design of the US position at the UNFCCC negotiations, and the European industry contributed to block the adoption of the carbon/energy tax, thus crippling EU attempts to perform a leadership role at the UNFCCC negotiations. Thus, the industry had a high capacity for blocking progressive national climate negotiating positions, and the coalition maintained a reactive strategy.

With the adoption of the Kyoto Protocol five years later, however, the situation had changed. The Kyoto Protocol went far beyond the industry's interests, particularly in terms of the inclusion of specific reduction targets and timetables, and the exclusion of commitments for developing countries. During this phase the influence of the industry on the regime development was (temporarily or permanently) reduced and the regime had a stronger impact on the strategies of the industry – at least in Europe. The agreement was adopted quite contrary to the interests of the fossil-fuel industry and despite vigorous attempts, particularly by the US-based companies, to stop it. The European industry – including Shell – revised its strategy in response to a situation where its capacity to influence the climate position of the EU was significantly reduced. As we have seen, the European industry moved towards a more proactive strategy after the adoption of the Kyoto Protocol.

In the US, the Clinton–Gore administration excluded the fossil-fuel lobby from the decision-making process preceding and during the Kyoto negotiations. The exclusion was temporary, however, since the fossil-fuel industry maintained its influence through the Byrd/Hagel resolution and its alliance with Congress. This contributes to explaining why the US industry did not shift its strategy towards a more proactive approach in response to the adoption of the Kyoto Protocol. For the agreement to be ratified by the US and become US law, it had to acquire support by a two-thirds majority in Congress. Thus, ExxonMobil was convinced that the US would never ratify the Kyoto Protocol, maintained its reactive stance and escalated its mobilisation against the agreement. Thus the influence of the US-based industry remained unaltered, despite the temporary setback during the Kyoto negotiations, and has largely maintained its reactive approach to the climate problem.

Until quite recently, most observers considered an international climate regime without the participation of the US to be highly unrealistic. With its 25 per cent share of global GHG emissions, the US is a pivotal actor in global efforts to abate the climate problem. Thus, the US fossil-fuel lobby had every reason to believe that the country's withdrawal from the Kyoto agreement would sink the international effort to regulate GHG emissions. With the EU's apparent success in mobilising sufficient support for the Kyoto agreement to enter into force even without US participation, the US fossil-fuel lobby did not sink the agreement, although its influence has affected the inclusiveness of the regime.

After Kyoto, the fossil-fuel lobby is no longer a cohesive interest group in the negotiations, and European- and US-based companies face significantly different regulatory frameworks. In essence, the IR model is supported by the observation that the climate strategies of Shell (and the European fossil-fuel industry) and ExxonMobil (and the US fossil-fuel industry) correspond to their respective influence on the climate regime.

As it turned out, the adoption of the Kyoto Protocol was a precursor to an emerging divide between Europe and the US, in terms of both policy measures and corporate strategy. Regimes may start out with a narrow scope and lenient commitments, but as long as they include feedback mechanisms around which actor expectations converge, a dynamic development is encouraged. As discussed in chapter 2, we believe the combination of the 'snowball effect' at the regime level and the sovereignty principle is a key mechanism whereby international regimes can induce change in corporate strategies. The principle of sovereignty provides states with significantly more rights than large corporations. In particular, sovereign states have the right to refrain from international agreements that they have not given their consent to. Corporations do not have this right. Thus, when states choose to join international regimes that become progressively more 'demanding', the decision environment for large corporations also changes. With the EU ratification of the Kyoto Protocol, the framework conditions for the European fossil-fuel lobby became different from those of the US-based industry.

Our analysis has focused on three causal pathways through which the international regime may affect corporate strategy choice: knowledge, pressure and opportunity. All of these mechanisms seem to have been in operation. First, the knowledge-based pathway is linked to the industry's acceptance of and response to the knowledge base provided by the IPCC. While the US fossil-fuel lobby has based its reactive position on a refutation of IPCC conclusions, the European oil industry accepts these conclusions as authoritative climate knowledge. As we have seen, the IPCC is organised in a manner that ensures equal access to the process for all stakeholders. The main difference between the European and the US-based industries, therefore, lies in their receptiveness to the knowledge provided. While ExxonMobil has its own research capacity on climate science, Shell does not and is thus more dependent upon IPCC conclusions. It is important to note, however, that factors linked to the gradually evolving knowledge base cannot explain the rapid change that took place in European industry strategies. That is, there were no breakthroughs in the knowledge base immediately preceding, for instance, Shell's strategy shift that could have induced or persuaded Shell to move from a reactive to a proactive climate approach.

Second, we have seen that the climate policy of the EU exerted an increasingly strong regulatory pressure. In chapter 2, we assumed that a strong regime would promote proactive corporate responses by shaping mutual expectations about the need for future regulation. With the 1998 burden-sharing agreement, both the EU and each member state have legally binding targets and share the responsibility for meeting these. The ECCP, combined with a monitoring mechanism that has been in place since 1993, has further promoted EU implementation of its Kyoto commitments. Within this regulatory framework, it was increasingly risky for the European industry to adopt a reactive response.

Third, this development was reinforced by the new opportunities the EU climate regime ensured in its policies for promoting renewable energy resources and in the development of an internal scheme for emissions trading. As noted above, there is a noteworthy correspondence, first, between the EU's adoption of the 12 per cent target on renewable energy sources and Shell's establishment of its fifth core business area – Shell International Renewables – and, second, between the Green Paper issued by the Commission on the design of an emissions trading system and Shell's launching of its own internal emissions trading system (STEPS).

In Europe, therefore, climate policies were characterised by a common understanding of the climate problem upon which increasingly authoritative and ambitious measures to implement EU Kyoto commitments were based. In addition, new business opportunities were generated. Concurrent with these developments we see a remarkably quick transformation of most, if not all, European-based oil companies from their reactive stance towards a more proactive approach from which they explicitly express their support for the Kyoto Protocol.

It is also important to note, however, that these factors operate in combination with equally authoritative regulatory measures at the national level, in the companies' home-base countries. The home-base country of the Shell Group, for instance, the Netherlands, has adopted an ambitious climate policy since the early 1990s and was among the first countries to ratify the Kyoto Protocol. The Netherlands, moreover, has also adopted a progressive policy to increase the share of renewable energy sources in national energy consumption. Thus, Shell has been subjected to 'double exposure'. With limited options to influence and block progressive and ambitious climate policies and negotiating positions at either level, this situation has significant implications for Shell's strategy choice.

In sum, our analysis indicates that the extent to which corporate actors can influence an international regime indeed has an impact on their strategy choice. A high degree of influence tends to pull the industry towards a reactive approach, while a low degree of influence tends to pull the industry towards a more proactive response at least when combined with regime 'maturation'. Moreover, our analysis indicates that in a situation where the industry has limited influence, international regime features linked to the provision of a common knowledge base, regulatory pressure and new market opportunities seem to constitute important factors for understanding the dynamics of corporate strategy choice.

Notes

1 Personal communication with Phillip Cooney, American Petroleum Institute, and William O'Keefe, Solutions Consulting, Washington, DC, March 2000.

2 *International Environmental Reporter*, 30 July 1991, p. 421: 'Oil industry presents paper to Commission opposing creation of tax on energy sources'.

3 Personal communication with Glenn Kelly and Eric Hold in the Global Climate Coalition, Washinton, DC, March 2000.

4 *Planet Ark*, 9 November 2000, 'Global warming business group cools its message'.

5 Personal communication with Brian P. Flannery and Gary F. Ehlig, ExxonMobil Corporation, Irving, Texas, March 2000; Phillip Cooney, American Petroleum Institute, and William O'Keefe, Solutions Consulting, Washington, DC, March 2000; Glenn Kelley and Eric Hold, Global Climate Coalition, Washington, DC, March 2000.

6 Personal communication with representatives of ExxonMobil, API, GCC and the Pew Centre, in Dallas, Texas, and Washington, DC, March 2000.

7 Personal communication with Brian P. Flannery and Gary F. Ehlig, ExxonMobil Corporation, Irving, Texas, March 2000.

8 Personal communication with Phillip Cooney, American Petroleum Institute, and William O'Keefe, Solutions Consulting, Washington, DC, March 2000.

9 *ENDS Environment Daily*, 1152, Wednesday 6 February 2002, pp. 2–3, 'Anti-Kyoto business lobby "Deactivates"'.

10 Climate Action Network (CAN), November 2000, COP-6 position paper. Source: www.climnet.org.

11 Just as the fossil-fuel lobby allied with OPEC countries, ENGOs allied with the parties most vulnerable to climate change. For example, Greenpeace and the Foundation for International Law and Development (FIELD) have strengthened the position of the Alliance of Small Island States (AOSIS).

12 FoE, 1992, 'The carbon/energy tax and industry: exploding the myths'.

13 One important weakness referred to is the lack of targets. See 'Emissions trading in the EU: let's see some targets!', 20 December 2001. Source: www.climnet.org.

14 Personal communication with Phillip Cooney, American Petroleum Institute, and William O'Keefe, Solutions Consulting, Washington, DC, March 2000.

15 Personal communication with Phillip Cooney, American Petroleum Insitute, and William O'Keefe, Solutions Consulting, Washington, DC 2000.

16 Personal communication with Phillip Cooney, American Petroleum Institute, and William O'Keefe, Solutions Consulting, Washington, DC, March 2000.

17 The most important were: (1) the incorporation of six GHGs instead of three; (2) inclusion of sinks; (3) multi-year targets instead of single-year; and (4) a market-based approach allowing for emissions trading.

18 Interview with Eileen Claussen and Sally C. Ericsson, the Pew Center on Climate Change, Washington, DC, March 2000.

19 E. Masood, 'Climate report "Subject to scientific cleansing"', *Nature*, 381, 13 June 1996, p. 546.

20 *Wall Street Journal*, 12 June 1996. See also *Bulletin of American Meteorological Society*, 77: 9, September 1996, pp. 1961–6, where all the correspondence in The *Wall Street Journal* on this occasion is reprinted. For a more detailed account of this incident, see Skodvin, 2000b: 215–19.

21 E. Masood, 'Climate report "subject to scientific cleansing"', *Nature*, 381, 13 June 1996, p. 546.

22 E. Masood, 'Companies cool to tactics of global warming lobby', *Nature*, 383, 10 October 1996, p. 470.

23 Personal communication with Brian P. Flannery and Gary F. Ehlig, ExxonMobil Corporation, Irving, Texas, March 2000; Gerry Matthews, Shell International, Washington, DC, March 2000.

24 Personal communication with Brian P. Flannery and Gary F. Ehlig, ExxonMobil Corporation, Irving, Texas, March 2000.

25 Personal communication with Gerry Matthews, Shell International, Washington, DC, March 2000.

26 'Remarks by President Bush on global climate change', 11 June 2001. Source: www.state.gov/g/oes/rls/rm/4149.htm (accessed 8 March 2002).

27 *Planet Ark*, 5 April 2001: 'ANALYSIS: Bush's climate stance cheers scientific sceptics'.

28 *Los Angeles Times*, 4 April 2002: 'Charges fly over science panel pick'. Source: www.latimes.com (accessed 8 April 2002); *Guardian*, 5 April 2002: 'Oil giant bids to replace climate expert'. Source: www.guardian.co.uk (accessed 5 April 2002).

29 Personal communication with Phillip Cooney, American Petroleum Institute, and William O'Keefe, Solutions Consulting, Washington, DC, March 2000.

30 Interview with Barend van Engelenburg, Ministry of Housing, Spatial Planning and the Environment, The Hague, November 2000.

31 *Planet Ark*, 4 March 2002: 'Dutch set pace on Kyoto treaty ratification'.

32 European Commission, Briefing Paper, 6 July 2001: 'EU position for the Bonn conference on climate change 19–27 July 2001.' Source: europa.eu.int/comm/environment/climat/eupositions.htm.

33 'Community legislation in force, Document 300D0646: Decision No 646/2000/EC of the European Parliament and the Council of 28 February 2000 adopting a multiannual programme for the promotion of renewable energy sources in the Community (Alterner) (1998 to 2002)'. Source: europa.eu.int/eur-lex/en/lif/dat/2000/en_300D0646. html (accessed 25 February 2002).

34 EU Commission, 14 May 2001: 'Green Paper on greenhouse gas emissions trading within the European Union. Summary of submissions. Part I: Non-governmental submissions.'

35 Interview with Valèrie Callaud, EUROPIA, Brussels, November 2000.

36 *Planet Ark*, 22 February 2002, 'Emissions trading systems developing as patchwork'.

37 ExxonMobil has the highest share of representation in the Secretariat General and the Board of Directors in EUROPIA, and leads EUROPIA sections on air quality, industrial sites and environment.

7
Concluding remarks

How different are the climate strategies adopted by major oil companies? Why do they choose different strategies, and what triggers changes? In addressing these questions, we have made an effort to identify the key conditions determining the climate strategies of large oil companies. The oil industry makes a living from the main sources of GHG emissions and exercises significant political influence at both national and international levels. A natural strategy for oil companies has been to eschew climate-change regulation. In the early 1990s, the oil industry was united in its opposition to binding climate targets. A precondition for a viable climate regime is thus a change in the strategies of large multinational oil companies. Governments depend on the active or reluctant cooperation of this industry for mitigating climate change. The identification of conditions that determine how the climate strategies of major oil companies are formed may thus provide us with knowledge about the extent to which and how corporate support for a viable climate policy can be stimulated and corporate resistance overcome.

To address the research questions and move towards a better understanding of factors explaining changes and differences in corporate climate strategies, we have chosen to focus on three major oil companies in this study: ExxonMobil, the Shell Group and Statoil. Crudely put, these companies share the same core aim of selling as much oil and gas as possible at the highest possible price and the lowest possible cost within the same global market. The business opportunities and challenges offered by regulatory measures to curb GHG emissions would thus apparently be the same for these companies. Nevertheless, ExxonMobil, Shell and Statoil have chosen three different strategies to meet the challenges offered by the problem of climate change. Shell has chosen a proactive strategy, ExxonMobil a reactive one, while the climate strategy adopted by Statoil can be placed in between as 'intermediate'. A systematic scrutiny of this puzzle represents a good opportunity to generate a better understanding of the sources of corporate climate strategies.

The point of departure for the analysis in this book is the sharp contrast between the important role played by the oil industry and the lack of analytical frameworks within social science climate research for studying corporate actors. Systematic case studies of major companies in other issue areas are also short in supply, even though a wide range of global environmental problems has been linked to the worldwide operations of multinational corporations. In this chapter, we will first recapitulate the analytical framework developed and applied in this analysis, before we sum up our empirical findings and their implications for analysis and policy. We believe that the analytical framework developed here may be applicable also for analysing other issue areas in which large corporations play an important role, such as ozone depletion, genetically modified organisms and emissions of hazardous substances.

The analytical approach

The explanatory focus in this study has been corporate climate strategies. We have based our comparison of the three companies on a continuum from reactive to proactive strategies, with an emphasis on what corporations actually do. This emphasis is extremely important, since the rhetoric used by the companies can deviate significantly from their actual operations. In the absence of reliable and comparable time-series data on GHG emissions, we have assessed the companies' climate strategies on the basis of a set of indicators that may provide us with an indication of the kind of climate policy futures the companies are preparing for: (1) their acknowledgement of the problem; (2) their positions on the Kyoto Protocol; (3) their GHG emissions reduction targets and measures; and (4) the extent of reorientation in their core business areas.

To explain differences and change in corporate climate strategies, three perspectives or 'models' were developed. These differ in their relative emphasis on explaining differences versus change in corporate strategies, but all three models are relevant for understanding both dimensions. The *Corporate Actor (CA)* model is based on the assumption that strategies vary as a result of differences between the companies themselves. This model is based on the business environmental management literature and consists of three core variables with high relevance for climate-change strategies: the environmental risk associated with the company's activities; the company's environmental reputation (measured in terms of the company's experience with negative public scrutiny); and the company's capacity for organisational learning. In addition, a number of other company-specific factors, such as leadership, capital and human resource availability, corporate tradition and ownership structures, have been included in the discussion as potentially moderating. The effects of these moderating factors are, however, extremely difficult to determine analytically and measure empirically within this specific context. At a very general level, the proposition derived from the CA model predicted that variation in company-specific features would lead to a similar variation in climate strategies. More specifically, we assumed that a low level of environmental risk associated with the company's activities, experience with negative public scrutiny affecting the company's environmental reputation, and high organisational learning capacity would lead to a proactive strategy on climate change. Likewise, we assumed that the converse – i.e. high environmental risk, no negative public scrutiny, and a low capacity for organisational learning – would lead to a reactive climate strategy.

The second perspective – the *Domestic Politics (DP)* model – postulates that differences in climate strategy can mainly be explained by differences in the national political contexts of the companies rather than in the companies themselves. This model is based on theories of state–society relationships applied to multinational corporations. The DP model is based on the assumption that even global companies are strongly tied to a home-base country in which they have their historical roots, located their headquarters and concentrated much of their activities. The national political context of the companies' home-base countries is likely to affect corporate climate strategy through the following core factors: social demand for environmental quality; governmental supply of climate policy; and the nature of the political institutions linking demand and supply, here understood as state–industry relationships. On the basis of the DP model, we assumed that at a very general level, different political contexts would lead to different climate strategies. More specifically, we assumed that a strong social demand for environmental protection measures, supply of an ambitious climate policy, and cooperative and consensus-oriented political institutions would promote proactive corporate strategies.

The last model developed for understanding corporate climate strategies was labelled the *International Regime (IR)* model. This is based on theories of international regimes (particularly regime effectiveness approaches) and is particularly suitable for understanding change in corporate climate strategy. The IR model departs from the focus on single companies and their home-base countries and directs attention to corporate alliances operating within the scope of international institutions liable to change. The main premise of this model is that the direction of influence – that is, the influence of industry on regimes versus the influence of regimes on industry – is likely to affect corporate strategy. If a reactive industry is capable of determining regime commitments, persistent reactive corporate strategies can be expected. Conversely, if the regime progresses beyond industry opposition, we can expect that high regime pressure, in terms of stringent joint commitments, provision of good market opportunities by the regime, and provision of a common knowledge base to which corporations are receptive, will lead to a proactive strategy among companies.

The three models presented above are linked to a more general debate on the limitations and possibilities of *governance* in the present world system. The CA model posits that the sources of corporate climate strategy can be found within the companies themselves. This assumption suggests that differences between the companies themselves may be more important than differences in political context for understanding corporate climate strategies. In contrast, the DP model suggests that differences in the political and societal contexts of the corporations' home-base countries may affect corporate climate strategy even though company-specific features remain constant. And the IR model suggests that multinational companies can be influenced by international environmental institutions.

The implications derived from each of the three models presented above rest on crude simplifications of realities. Multinational oil companies certainly do not operate in a societal and political vacuum, and even the distinction between 'company-specific' and contextual factors is not always clear cut. We have pointed to a number of instances of interplay and causal complexity between and even within the models. Before these analytical challenges are discussed any further, we will briefly summarise the main findings of the various chapters.

Empirical findings of individual chapters

The aim of chapter 3 was to assess the climate strategies of ExxonMobil, Shell and Statoil in relative terms according to the four indicators. We found that while ExxonMobil acknowledges the climate problem as a 'legitimate concern', it does not accept the current status of climate science as a sufficient basis for regulatory action. In contrast, Shell and Statoil both accept climate change as a real problem that requires collective action on a global scale. The companies' positions on the Kyoto Protocol vary accordingly: ExxonMobil opposes the Protocol, while Shell and Statoil support it. It is important to note that with regard to ExxonMobil and Shell, the difference in their positions goes far beyond a difference in rhetoric: ExxonMobil has vigorously opposed the Protocol, while Shell has taken a number of voluntary actions to support it. These include the adoption of voluntary and ambitious emissions reduction targets for its own operations, combined with reduction measures such as the establishment of an internal emissions trading system. Statoil, on the other hand, has adopted a rather ambiguous reduction target mainly linked to technological innovation. ExxonMobil has not adopted any emissions reduction targets or measures for its own operations. While the difference between ExxonMobil and Shell is perhaps the most striking, the difference between Shell and Statoil also becomes clear when we look at long-term commitments in the form of a reorientation in core business areas: Shell has divested its coal assets, changed its decision routines to include carbon costs for future projects, and established renewable energy resources as its fifth core business area; Statoil has not undertaken similar actions.

Taken together, these indicators thus show that the three oil companies under scrutiny here have chosen three significantly different climate strategies: ExxonMobil has adopted a reactive strategy, Shell a proactive one, and Statoil an 'intermediate' one that can be placed in between ExxonMobil and Shell. In the case of Statoil, there seems to have been a gradual process of incremental change (although not a very large one) from the early 1990s. Shell, on the other hand, rather abruptly changed from a reactive to a proactive strategy around 1997/1998.

The aim of chapter 4 was to assess the fruitfulness of the CA model for explaining differences in corporate climate strategy. The main conclusion, based on the three core dimensions of the model, was that the similarities in company-specific factors between the three companies are much more striking than the differences. All three companies face significant environmental risk, they have all had experience with negative public scrutiny, and they have all developed a certain level of learning capacity. As expected, the CA model was unable to explain the changes in corporate strategy of the three companies under scrutiny here, because the model variables did not systematically change prior to the observed changes in climate strategy.

This being said, the CA model cannot be deemed irrelevant for understanding corporate climate strategy. The moderate differences observed between the companies did pull in the directions that could be expected on the basis of the model. For instance, while all the companies face a significant environmental risk associated with their operations, ExxonMobil is the most carbon intensive of the three and thus the one most vulnerable to GHG regulation. According to the CA model, this would increase the likelihood of a reactive response. Similarly, Shell has a higher capacity for organisational learning than the two other companies particularly because of the company's long tradition of scenario development, representing a systematic monitoring of environmental, societal and political trends. According to the CA model, Shell's higher learning capacity increases the likelihood that the company will choose a proactive response to climate change. Nevertheless, while the CA model helps to explain differences in the climate strategies observed, the analysis in chapter 4 indicates that the overall explanatory power of the model is weak. To check the robustness of these conclusions, moreover, a number of additional company-specific factors suggested in the business environmental management literature were briefly scrutinised. These factors, however, did not significantly modify the general conclusions of the chapter.

The aim of chapter 5 was to assess the merits of the DP model. This model proposes that the political contexts characterising the home-base countries of ExxonMobil, Shell and Statoil (the US, the Netherlands and Norway respectively) would explain those companies' climate strategies. The analysis shows that, judged on the merits of pattern-matching, the propositions derived from the DP model gain strong empirical support. All three cases match the expectations derived from this model: Shell has its backbone in the Netherlands and has adopted a proactive strategy. This outcome matches well with the observations that the Dutch population has expressed the strongest demand for environmental and climate policy, the Netherlands has adopted the most ambitious climate policy in terms of targets and policy instruments, and Dutch political institutions are founded on a tradition marked by cooperation and consensus-seeking between the state and industry. This pattern appears robust even when controlled for the fact that Shell International is based in the UK. In short, a positive interplay between a social demand for environmental protection measures, governmental supply of climate policy and the nature of political institutions linking demand and supply serves to provide Shell with both pressures and opportunities for adopting a proactive strategy.

Conversely, the reactive strategy adopted by ExxonMobil has evolved within a political context characterised by a relatively weak demand for climate measures, low governmental supply of climate policy, and political institutions linking demand and supply characterised by conflict: in relative terms, North Americans express less concern about climate change than the Dutch, and US climate policy has been weak since the adoption of the UNFCCC. The political institutions channelling industry–state influence are characterised as adversarial in spite of recent efforts to develop a more cooperative approach. In essence, therefore, a negative interplay characterises the domestic political factors in the US. This serves to expose ExxonMobil to a low regulatory pressure and to provide few new market opportunities, and thus induces a reactive corporate response. Norway and Statoil can be placed in between these extremes along the core dimensions of the DP model. Norway's social demand has fluctuated significantly during the 1990s and its climate policy has been ambiguous. Thus, we also have no empirical evidence to suggest that state-owned companies are more inclined to choose a climate policy in line with the policy of their government owners.

The causal mechanisms underlying these patterns are not, however, entirely persuasive. The limitations of the DP model clearly surface when it is confronted with changes (rather than mere differences) in the core variables. While the key factors of the DP model provide a strong basis for explaining differences in strategy choice between the companies, they do not provide satisfactory explanations to the question of why changes in strategy choice occurred or did not occur at specific points in time. For instance, changes in the Netherlands do not appear sufficiently significant to explain why Shell's turnabout from a reactive to a proactive position in 1997/1998 came when it did. Similarly, there were changes in the US political context – notably the signing of the Kyoto Protocol in 1997 – that would seem sufficiently dramatic to induce a change in ExxonMobil's climate position, and the DP model is incapable of explaining why such a change did not occur. The analysis thus shows, as expected, that the DP model is significantly more capable of explaining *differences* in corporate strategy choice than *changes* in corporate climate strategy over time.

In chapter 6, we assessed the explanatory power of the IR model. The propositions of this model were generally supported by the observations that the climate strategy adopted by European fossil-fuel industries (including Shell) and US fossil-fuel industries (including ExxonMobil) correspond with their respective influence on the climate regime. US fossil-fuel industries have maintained a high level of influence on the climate regime mainly through domestic channels because of the pivotal position of the US at the international level. The Byrd/Hagel resolution adopted by the Senate before the Kyoto Protocol led ExxonMobil to believe that the Kyoto Protocol would never be ratified by the US

Senate and thus probably not be ratified at all. The company has, therefore, never seen the need for developing a 'plan B' in the case of ratification. In essence, the influence of ExxonMobil and the US fossil-fuel lobby on US foreign climate policies and consequently the international climate regime contributes to our understanding of why ExxonMobil did not soften its position after the US adoption of the Kyoto Protocol.

In contrast, European fossil-fuel industries experienced a loss in influence after the adoption of the Kyoto Protocol. The EU proved to be determined to ratify the Protocol and adopt a climate-change programme for meeting its commitment. The international climate commitments confronting the European fossil-fuel industry thus became different from those confronting the US-based companies. This led to the dissolution of the fossil-fuel lobby as a unified group in the climate negotiations. Our analysis shows that in tandem with the European climate regime that emerged, the international climate regime affects corporate climate strategies (mainly) in Europe through three causal pathways. First, it affects corporate strategies by exerting regulatory pressure, particularly through the Kyoto Protocol and the ECCP. Second, it affects corporate strategies by providing new market opportunities at the EU level, particularly in the form of a new market for renewables and a market for CO_2 emissions through a European scheme for emissions trading. Third, it affects corporate strategies by providing a common knowledge base. The perception of this knowledge base did, however, vary between European and US-based industries because of differences in their receptiveness to the knowledge provided. There were no major breakthroughs in the knowledge base that can explain the change witnessed in Shell's climate strategy.

Analytical implications

The first observation is that variance in the domestic political context of the companies' home-base countries is more important for explaining differences in corporate climate strategy than are company-specific factors. This finding is somewhat surprising in the light of the business environmental management literature, which is specifically intended to represent the real-world phenomena that shape corporate environmental strategy. One main limitation of this literature lies in its strong emphasis on company-specific factors in isolation. In this study, similarities rather than differences dominate the picture of company-specific features. Even the 'hard' economic indicators applied here, like core business areas, exploration and production volume and resource reserves, did not add much to our understanding of why the three companies had chosen three different climate strategies. This approach, therefore, appears less capable of explaining (differences in) corporate strategy choice when the companies in question operate in the same business area and bear similar company-specific features.

At least two notes of caution are in order with regard to the limitations of the business environmental management literature. First, we should bear in mind that many of the company-specific factors pointed to in the business environmental management literature are extremely difficult to determine and measure in a comparative perspective. Many of them – for instance, corporate leadership and shareholder pressure – can affect climate strategies in rather unpredictable ways. Second, the CA model applied here has been constructed and simplified on the basis of more general approaches. For example, the business environmental management literature does recognise 'political/legal' environments as sources of corporate choice. But such factors tend to be narrowly treated as 'governmental regulation' only. This study shows that governmental regulation represents only one, and not necessarily the most important, domestic political factor that affects corporate strategies.

The relatively high explanatory power of the DP model is equally surprising, but for a different reason. This model is based on theories of society–state relationships and thus not specifically designed to explain corporate environmental strategy. The DP model has been developed to understand political rather than corporate decision-making. In democratic societies, politicians are, crudely put, concerned with votes while corporate leaders are concerned with profits. This means that politicians and corporate decision-makers are sensitive to differences and changes in social demands for different reasons. Politicians respond to please their electorate, while corporate decision-makers respond to please their consumers – although the link between social demand and a change in voting behaviour can be seen as more direct than that link between social demand and a change in consumer behaviour. Social demand is, however, important for companies for a second reason: corporate strategy can be affected indirectly to the extent that social demand affects the level of governmental regulation.[1]

The supply side of the DP model proved directly applicable to understanding corporate strategy, since it focuses on domestic implementation. Degree of implementation is based on the assumption that target groups, such as companies, 'control' the behaviour that has to be modified in order to comply with higher-order goals. This line of reasoning links approaches to domestic politics to corporate strategies in at least two ways. First, the DP approach is explicitly concerned with how regulatory policies affect the strategies of target groups. Second, the model highlights the extent to which governments control society, including powerful target groups. The latter dimension proved to be important because it captures the consequences of different political institutions, i.e. the way in which governments interact with industry. Moreover, this study indicates that the relative control between oil companies (and other fossil-fuel industries) and governments also depends on international regimes, and varies over time and between countries. All in all, the DP model proved to be helpful as a tool for better understanding the importance of 'nationality' that is only hinted at in the literature on business environmental management.

The second general observation is that the IR model proved to be particularly suited for explaining change (or lack of change) in (collective) corporate strategies over time. This is not surprising given the main purpose of this model. Although the CA and DP models also carry a potential for explaining changes in corporate strategy by emphasising changes in company-specific features and domestic political context, they fall short because they are unable to capture the dynamics of the international arena. And if the business environmental management literature merely tends to downplay domestic political context, the influence of international institutions is virtually absent. This is a serious shortcoming given the large number of international environmental agreements that affect corporate decisions. The merit of the IR model lies in the link between international or global institutions and multinational or global corporations. Changes in the scope and stringency of international institutions can produce significant collective changes in business strategies. This observation is also an argument for realising the long-lasting ambition of regime scholars to focus more systematically on non-state actors and international institutions, particularly powerful industry groups.

Even though our three models differ in explanatory power and main focus, the third observation is that all three models are complementary in nature. Each has contributed something that the others have not. Even the CA model has provided unique insight into this case, in the sense of being particularly well suited for understanding how different corporations within the same branch interpret and perceive the outside world differently.[2]

The last observation is related to the complementary nature of the three models: an important key to understanding corporate climate strategies lies to some extent in the interplay and casual complexities within and between the models. To recapitulate from chapter 2: causal complexity refers to situations where the *direction* of influence depends on the value of another variable, while interplay concerns situations where the *effect*, or the explanatory power, of one independent variable is conditioned by the value of another variable. One example of causal complexity within the models is the effect of learning capacity on corporate climate strategies. The way oil companies perceive initiatives to facilitate increased use of renewables may depend on past experience with investments in this field. Companies that have had a positive experience are liable to see such initiatives as an opportunity, while companies that have suffered negative experiences may perceive the same situation as a threat. An example of interplay can be seen in the DP model. A strong social demand for environmental quality and governmental supply of environmental policy pull in the same direction, but the level of supply is (partly) conditioned by social demand: a company based in a country characterised by a relatively strong social demand for environmental quality is also likely to be exposed to stringent environmental regulation.

At the interface of the three models, there is first a relationship between the CA and the DP models. A company's likelihood of being exposed to negative public scrutiny depends on the societal context in which the company operates. Even though the activities undertaken may be the result of strategic decisions, the consequences for the company also depend on external political factors. Shell's experience with the Brent Spar incident illustrates this point: on this occasion, Shell's decision to dump the redundant platform was based on considerations of marine pollution and economy, and the company did not anticipate the public outcry the decision caused, which ultimately forced Shell to reconsider and withdraw its decision to dump the platform in the ocean. Another close relationship between these models is related to learning capacity, particularly the extent to which environmental trends are monitored systematically. As we have seen, learning capacity is at one level related to previous activities linked to the company itself and represents in this sense a company-specific factor. At another level, however, learning relates to the societal and political context in which the company operates. In addition to its capacity to foresee future trends, what the company sees also depends upon where it looks. Shell, for example, looks mainly to the Netherlands and the EU and foresees, by means of scenarios, a future where the Kyoto Protocol is ratified (either fully by the OECD, or partially by the EU) and where there is a stronger demand for cleaner energy in the form of renewables. Statoil looks mainly to Norway and Scandinavia and perceives natural gas as the bridge to alternative energy carriers. And ExxonMobil directs most of its attention to the US and does not see any significant changes in the demand for fossil fuels, at least not in the short term. These links between companies and contexts are further reinforced by the relative importance of their main markets: the US market is more important to ExxonMobil than is the European market, while the European market is more important to Shell than is the US market. The Norwegian/Scandinavian market is most important to Statoil.

Just as company-specific features interact with domestic contexts, so domestic contexts interact with international institutions. In this study, regime-specific factors have tended to operate in combination with factors linked to the companies' home-base countries at two levels. The first is related to the channels of corporate influence. Because of the pivotal role of the US in international climate policy, US fossil-fuel companies have largely relied on domestic US influence to put the Kyoto Protocol out of action. The other combination is related to pressures and opportunities: Shell has been subjected to 'double exposure' from the Netherlands and the EU with regard to *both* new markets for renewables and an, in relative terms, ambitious climate policy. An influential international regime does not, however, of necessity produce similarity in relevant domestic political factors. Even though countries are subjected to similar international climate commitments, their responses in terms of implementation and compliance can and actually do vary widely. One prominent example is the difference in climate policies between Norway and the Netherlands.

We have also identified interaction between company-specific features and international knowledge production. Differences in in-house expertise and investment in R&D relevant to climate change did affect the companies' receptiveness to the knowledge provided by the IPCC.

The analysis shows that the combination of factors that maximises the likelihood of a proactive strategy occurs when a company associated with a high environmental risk, with a high capacity for learning and previous experience with negative public scrutiny operates within the framework of a home-base country characterised by a strong social demand for environmental quality, governmental supply of a relatively ambitious climate policy, and political institutions seeking consensus and cooperation with industry. It is important that the home-base country be committed to an international regime that rests on a common knowledge base and produces increasingly strong joint commitments and market opportunities.

Alternative or supplementary approaches

Even though this study has brought us closer to an understanding of the significant differences and changes in corporate climate strategies that have occurred, there is still room for other explanatory perspectives and approaches. Any judgement of the relative fruitfulness of different models and approaches should be based on explicit criteria such as generality, conclusiveness, validity and parsimony (Underdal, 1984; Skjærseth, 1995). We may argue that one model is more fruitful for empirical analysis than another if it can explain the same patterns as alternative models plus something more, at equal operational costs.

In this study, the companies' climate strategies have mainly been analysed independent of each other. Climate strategies have been seen as a function of corporate, domestic and international context factors. This approach is well suited to identifying the mechanisms linking corporate choice and its institutional context. It is less suited, however, to capturing how companies relate to each other. For example, BP was the first company to withdraw from the GCC. This 'first move' may have served to change Shell's business environment and may thus have influenced the company's turnabout from a reactive to a proactive climate strategy in 1997/1998. Further, we have indicated that a collective chain reaction took place from 1999 when a number of US-based companies withdrew from the GCC. Such group dynamics can be seen as 'critical mass phenomena' that can be analysed with a point of departure in Schelling's 'tipping model' (1978). This model was originally applied to understanding changes in settlement patterns triggered by changes in the proportion of minority immigrants within confined US neighbourhoods. Small proportional changes between minority and majority could take on an inner dynamic leading to significant shifts in settlement patterns.

The tipping model can be, and has been, applied generally to identify critical masses and aggregated outcomes. It seems reasonable to assume that company investments in renewable energy are influenced by the strategies adopted by other companies. Applied to corporations, the tipping model would predict that a change in strategy by one, or a critical number of companies, under certain conditions, could trigger a process of large-scale changes within and perhaps across industry sectors. An underlying mechanism here is 'the more the merrier': the more that join, the stronger incentives for others to join. The crucial question then is whether, and the extent to which, this mechanism applies to the oil industry and other fossil-fuel industries. The answer is not obvious since, for instance, the oil industry has begun to take on increasingly oligopolitistic characteristics (few companies and many buyers). The mega-mergers between Exxon and Mobil, BP and Amoco as well as the latest Texaco-Chevron and Philips-Conoco mergers are interesting in this perspective. The tipping model could thus add something new to our understanding of changes over time in corporate strategies, although it would be less suitable for explaining differences between companies in their strategy choice.

The analytical framework underlying this study is complex and not very parsimonious. One strategy to simplify the framework could be to merge the DP and IR models within the perspective of 'two-level' games (Putnam, 1988; Evans et al., 1993). At the national level, affected groups will promote their own interests by seeking to induce their government to adopt 'their' strategy, while politicians will seek to acquire power through the formation of coalitions between the groups. At the international level, the government will seek to maximise its possibilities of satisfying the same groups, while simultaneously minimising national costs of international commitments. If we replace 'groups' with the fossil-fuel industry and 'government' with the US, we will see that the notion of two-level games apparently fits very well with the present Bush–Cheney administration, but poorly with the climate policy of the previous Clinton–Gore administration. That administration apparently used costly international commitments to put domestic pressure on the fossil-fuel industry. As an alternative to the DP and IR models, the two-level games model could perhaps add some new insight, but probably at the expense of explanatory power. We should also bear in mind that the two-level games model has a rather narrow focus, since it is primarily concerned with the ratification of legally binding international commitments.

Policy implications

The study of oil companies and climate change can be seen as a particularly 'malign' case that represents a *critical test* of the limitations and opportunities of governance: these companies are among the largest and most powerful in the world and they face a strategic threat from climate change that might undermine their survival. If the climate strategies of such companies differ and change as a result of domestic and international governance, we will expect at least similar outcomes in less 'malign' cases. This indeed holds true in the sense that European industries generally are more proactive on climate change than are US industries. However, the causal complexities and interplay observed between company-specific features and political context factors imply that generalisation can be made only at a level confined within general causal patterns linking different levels of analysis. Thus, we can only offer conditional advice with regard to how domestic politics and international regimes can influence corporate strategies.

Let us illustrate this point by a crude replication of our analysis with other companies. The US is home to most oil majors, and we would, according to the relative explanatory power of our models, expect these companies to have adopted reactive strategies as well. On the one hand, this seems to be the case, since all major US-based oil companies, such as Texaco-Chevron and Philips-Conoco, oppose the Kyoto Protocol. On the other hand, Texaco has recently chosen a more proactive strategy than ExxonMobil. In 2000, Texaco acknowledged the climate-change problem and the company withdrew from the GCC. Likewise, while no major European oil company vigorously opposes the Kyoto Protocol, TotalFinaElf, for instance, is considered less proactive than BP and Shell.

Similar trends may be found within the auto industry. In 1998, the European, Japanese and Korean car makers adopted an agreement with the EU Commission to reduce CO_2 emissions from new passenger cars, while the US auto industry joined the US oil industry in its opposition to the Kyoto Protocol. Their opposition can be understood largely as a result of the environmental risk associated with their operations and the consumer demands in the US: the US market is most important to US auto companies and low fuel prices provide few incentives for consumers to care about fuel consumption. On the other hand, the US-based auto companies Ford and Chrysler left the GCC in 1999. Since then they have invested in low-emission technology and they have moved towards a stronger acknowledgement of the climate-change problem. One important driving force behind these recent changes in the climate strategies of US auto companies seems to be linked to mergers and joint ventures within the auto industry (Levy and Newell, 2000). Thus, the important point to make here is that the factors associated with the three models may interact differently in different cases and the driving forces for change are not necessarily available for manipulation by decision-makers. Instead of pushing this superficial replication any further, we will, with these caveats in mind, extract five concluding reflections with relevance for policy choices and with a view to the future. Notice that these reflections apply to climate policy in developed OECD countries only.

Major oil companies are governed by present political structures
Let us start with the good news from a governance perspective: this study supports the claim that multinational oil corporations operate within political structures that account for the differences between them. Such corporations are influenced by factors linked to their respective home-base countries, in which they have their historical roots, and have located their headquarters and their main operations. Social demand, governmental supply and political institutions channelling state–industry influence account for a significant proportion of the differences observed in climate strategies. These factors are to a varying extent available for manipulation by policy-makers.

Social demand for environmental quality represents an input to policy rather than the other way around. Social environmental attitudes can be influenced by the green movement, but are largely out of reach for policy-makers. Policy-makers can, however, support ENGOs. For instance, the EU Commission established the EEB as a counterweight to the business lobby. There have also been deliberate political efforts to influence social demand directly. The Clinton–Gore administration launched a number of US governmental initiatives aimed to increase public awareness of climate change. On the other hand, social demand is also stimulated by factors beyond the range of governance. Experience from other issue areas such as the Chernobyl accident, acid rain, ozone depletion and marine pollution indicates that social demand is particularly sensitive to 'shocks and crises'. In 1988, global warming surfaced on the US political agenda mainly as a result of heat waves and drought. Similarly, recent floods in Europe have been directly linked to climate change. Such events are potentially powerful drivers, but largely unpredictable and out of political reach.

Even though social demand and industry interests provide important input to governments, governments have significant latitude for acting independent of these pressures. Even state-owned oil companies that enjoy privileged access to decision-makers can be governed against their interests. In 1991, the Norwegian government adopted a CO_2 tax in spite of strong opposition from Statoil, which at the time was fully owned by the Norwegian state. Climate policy in terms of clear targets and mandatory policy instruments sends a strong signal to industry.

The combination of regulatory pressure from 'above' and stimulation of new market opportunities from 'below' is crucial for providing companies with incentives to adopt a more proactive climate strategy. For instance, in February 2002, President George W. Bush revealed his alternative plan (to Kyoto) on GHG emissions. The emissions reduction targets set by the scheme are tied to US economic growth and give companies incentives to meet them. In this regard, the plan meets one important condition for inducing proactive corporate strategies: the provision of opportunities. On the other hand, however, the scheme has been argued to amount to no more than 'business as usual' (Menz, 2002) in addition to being voluntary, and is thus unlikely to provide sufficient pressure to induce a change in the climate strategies of companies like ExxonMobil.

Political institutions channelling state–industry influence are difficult to change. Such institutions tend to be rooted in regulatory styles embedded in national history and tradition. For example, one of the most distinctive features of the Netherlands was the extraordinarily deep division of social and political life on the basis of religion. Another distinctive feature was related to a political culture characterised by compromise and consensus between the social and political 'pillars' at the highest levels. Even though general national features should not be exaggerated within specific issue areas, the Netherlands still relies heavily on compromise and consensus-seeking in environmental and climate policy. This tradition can even be traced to the company level: until its reorganisation, Shell placed a strong emphasis on reaching decisions by consensus, leading to an unusually high level of internal discussion. The robust nature of political institutions is also visible in the US. The Clinton–Gore administration made an effort to develop a more consulting and cooperative approach between government and industry, but analysts still claim that the adversarial and legalistic approach remains essentially intact.

Climate policy can be co-opted by major oil companies

The bad news from a governance perspective is that major oil companies closely tied to a home-base country do not necessarily adapt to policies generated from above. In the US, the fossil-fuel lobby, with the GCC, the API and ExxonMobil in the lead, has at least to some extent determined its own regulatory context, particularly by blocking the passage of legislation that threatens it. Therefore US climate policies have been based on public voluntary programmes independent of different administrations with varying climate policy ambitions. Corporate influence has been determined by various factors including the nature of the political system and the structural and instrumental power of the industry itself. The Clinton–Gore administration tried to limit the influence of the oil industry by keeping 'big oil' at arm's length. However, US climate policy, as other policies, is developed within a political system based on the US Constitution. And the US Constitution severely restricts the freedom of action of the executive branch.

The first significant victory of the US fossil-fuel lobby dates back to 1993. The industry mobilised against President Clinton's BTU tax and was able to sink the tax proposal with the help of a PR campaign. The tax initiative was perceived as extremely provocative by the fossil-fuel industry, and the Clinton–Gore administration was forced to rely on public voluntary programmes. A government's anticipation of what is politically feasible may be as decisive for its climate policy as a company's anticipation of future regulation is for its strategy choice. The industry's second major victory was the adoption of the Byrd/Hagel resolution by an impressive 95–0 majority in the Senate before the Kyoto negotiations. Since the Kyoto Protocol went well beyond the conditions set in the Byrd/Hagel resolution, the Protocol was deemed 'dead on arrival' (before ratification in the Senate), independent of the upcoming presidential election in 2000. US fossil-fuel industries have also targeted social demands. A number of campaigns have aimed to show the high costs of regulating GHG emissions in the US, particularly in terms of the increase in the price of petrol such policies would cause. Even single companies like ExxonMobil aim their education programmes at downplaying the environmental problems of fossil-energy sources while simultaneously stressing the costs of alternative energy.

'Proactive' multinational companies can weaken industry opposition in host countries

A national authority cannot require a company operating in another country to comply with its climate policy even if it represents this company's home-base country. A multinational company, however, can require its branch offices around the world to comply with its corporate strategy. Shell conducts business in some 135 countries, and its climate strategy applies in principle to all its branch offices in these countries. For example, Shell has a large North American branch office: Shell Oil in Houston. Together with BP, Shell was a member of the GCC, which has directed and coordinated a massive opposition to US ratification of the Kyoto Protocol. When Shell (and BP) changed their climate strategies, they left the GCC and joined the US Pew Center, thus to some extent weakening the opposition to GHG regulation and strengthening the corporate alliance in favour of a more ambitious climate policy in the US. To counterbalance the negative publicity flowing from the exits, the GCC had to reorganise itself from company- to branch-based membership – thus excluding all remaining single-company members and, in effect, reintroducing Shell and BP as members of the coalition via their membership in the API, which still was a branch member. Eventually, the GCC was 'deactivated' in 2002 – after 13 years in operation. The GCC claimed that it had served its purpose after the US exit from the Kyoto Protocol, but the Protocol is still alive and future US participation cannot be excluded. The declining support within the business community triggered by the exits of the European oil majors seems to have contributed to the 'deactivation' of the GCC, thus weakening the US fossil-fuel lobby. Conversely, corporate membership in the Pew Centre has risen from 21 major US companies in 2000 to nearly 40, most of which are included in the Fortune 500.

'Reactive' multinational oil companies have limited influence in host countries
The mechanism pointed to above could in principle also apply the other way around: ExxonMobil implements its reactive climate strategy all over the world and could therefore also serve to weaken industry support of climate policy in its host countries. However, we have few observations indicating such an effect. First, the climate policy of the EU has progressed even though US companies are strongly represented in European business organisations. Second, all major US oil companies are members of EUROPIA – the major European petroleum organisation – which

has shifted its position from opposing to supporting the climate policy of the EU. ExxonMobil is a major member of EUROPIA, but the current climate position of this organisation is clearly not determined by the lowest common denominator. In fact, the climate position of EUROPIA reflects the strategies of Shell and BP. Third, ExxonMobil and other US oil companies have major operations in Norway and the Netherlands. ExxonMobil, or its European branch Esso, pays the Norwegian CO_2 tax as do other oil companies, and the company participated in the Dutch LTA to enhance energy efficiency. The only alternative for ExxonMobil would be to move GHG emissions-generating operations from these countries to developing countries or the US. On the other hand, Norwegian efforts to reduce emissions of NMVOC by means of a purely voluntary agreement with the oil industry failed at the last minute due to massive opposition from US oil companies. These examples indicate that *mandatory policy instruments* can effectively counter negative influence from large reactive oil majors. As noted in chapter 3, a company like ExxonMobil places a strong emphasis on being in compliance with environmental laws and regulation.

International regimes can facilitate a 'snowball effect' capable of changing corporate strategies
'Indeed, if an agreement cannot be crafted that gains the consent of major affected industries, there will likely be no agreement at all' (Levy, 1997: 56). The past decade of climate-change negotiations indicates that this statement is only partly true. In the run-up to the 1992 UNFCCC, all major oil companies vigorously opposed all mandatory regulation of GHG emissions, and the UNFCCC did not oblige the parties to undertake such actions. Since then, the climate regime has developed in spite of massive opposition from a negatively affected industry.

At COP-1 in 1995, the parties agreed to the 'Berlin Mandate', which declared that non-binding commitments for developed countries were inadequate and that no new commitments would be imposed on developing countries. Two years later, the Kyoto Protocol was signed in spite of significant opposition from most fossil-fuel companies. The Kyoto Protocol committed developed countries to binding GHG emissions reductions while developing countries were excluded from similar obligations. In 1998, COP-

4 adopted the Buenos Aires Plan of Action, which aimed to facilitate the implementation of the Kyoto Protocol by solving the remaining 'core issues'. These issues, including flexibility mechanisms and compliance, were due to be resolved at COP-6 in the Hague in 2000. But COP-6 failed to reach an agreement until discussions resumed in Bonn in 2001 (COP-6 bis). Meanwhile, the newly elected President George W. Bush rejected the Kyoto Protocol. The political agreement struck in Bonn was subsequently translated into legal and operational terms at COP-7 in Marrakech. The 'Marrakech Accords' paved the way to ratification of the Kyoto Protocol. This will enter into force when ratified by 55 countries representing at least 55 per cent of CO_2 emissions in 1990. At the time of writing, 89 states, representing 37.1 per cent of global GHG emissions in 1990, have ratified the Protocol.

The development of the climate regime has experienced significant setbacks that will seriously undermine its effectiveness. Perhaps more surprising, however, is the fact that the Kyoto Protocol is still alive. This observation conforms well with the notion that institutionalised international cooperation gathers momentum through a 'snowball' effect that generates positive feedback and facilitates further steps. The conditions that increase the probability of a dynamic regime development are partly regime-specific: the climate regime started out with a narrow scope, lenient commitments and institutional feedback mechanisms that have encouraged a dynamic development. In addition, leadership has proved important. The EU has acted as a leader by means of 'showing the way' in its external climate policy as well as serving as a 'subregime' for other actors internally. The core mechanism explaining how an international agreement can progress beyond the opposition of major affected industries and subsequently influence their strategies lies in the combination of the 'snowball' effect and the limited scope for (direct) corporate influence at the international level.

Epilogue

International climate policy is still in a state of flux a decade after the UNFCCC. The US and the EU have drifted apart at the governmental level. This development prevented any progress on climate change and renewable energy sources at the Earth Summit in Johannesburg in September 2002. Major multinational oil companies are split on the issue, but the seeds of change towards a constructive climate pathway may actually lie in the link between international institutions and major multinational companies on either side of the Atlantic. The Kyoto Protocol still represents the most potent political force that has affected, and will most probably continue to affect, US multinationals with significant activities in Europe. On the one hand, US multinationals operating in Europe cannot take advantage of emissions reductions in the US, since the Kyoto Protocol does not recognise emissions reductions achieved in non-party countries. US companies exporting to Kyoto-party countries may face trade restrictions in the absence of voluntary measures, and incentives for developing new technologies within the world's largest economic market will be limited. US companies in Europe can further be exposed to negative public attention and possibly consumer boycotts. On the other hand, the Kyoto Protocol can provide energy-intensive companies in the US with a competitive advantage if energy prices in Annex B countries increase. However, if the US does not re-enter the Kyoto Protocol, the cost of emissions entitlements is likely to be much lower than anticipated by the US fossil-fuel industry because of the potential oversupply of surplus entitlements (hot air) from Russia and the Ukraine.

How positive and negative consequences add up for US multinationals remains to be seen. But European oil giants have already made their imprint in the US, and US companies in Europe are warming to the Kyoto Protocol. Early in 2003, ExxonMobil decided voluntarily to report carbon emissions, and the company is now backing mandatory reporting as a first step towards targets on emissions reduction. Over time, pressure from corporate non-state actors may pull towards a convergence between the US and the EU. A joint statement made between Greenpeace and some 160 multinationals on climate change at the Johannesburg Summit may stand as a symbol of the significant changes that have occurred in the business community over the past decade: Greenpeace and the World Business Council jointly called upon governments to tackle climate change on the basis of the UNFCCC and the Kyoto Protocol.

Notes

1 The strength of this mechanism should, however, not be exaggerated. Governments are not merely, and perhaps not even primarily, driven by social demand. Moreover, this mechanism is also likely to be conditioned by other factors, such as electoral systems. For example, electoral systems based on proportional representation are more likely to increase sensitivity to 'green' social demand than are 'winner-takes-all' systems. This is so for at least two reasons. First, 'green parties' stand a better chance of being represented. Second, many small political parties are more likely to absorb new demands than few and large political parties.

2 Ten to fifteen years ago, a concern for research economy would have been a strong argument against any attempts to open the 'black box' of company-specific features. Even large corporations did not release much relevant environmental information. At present, there is a lot of information to be found in booklets, at public conferences and not least on the companies' own Internet sites.

Appendix: personal communication

Interviews, taking the form of informal conversations, have been carried out with the individuals listed below. The positions referred to were those held at the time of the interview listed. In Norway, we have had more regular contact with representatives for Statoil, for instance, through Statoil's Environmental Forum.

ExxonMobil
Brian P. Flannery, Science Strategy and Programs Manager, Safety, Health and Environment. Irving, Texas, March 2000.
Gary F. Ehlig, Senior Advisor, Public Affairs Department. Irving, Texas, March 2000.
Guiseppe De Palma, Vice-President, European Union Affairs. Brussels, November 2000.

Shell International
Gerry Matthews, Advisor, Group Policy Development and External Affairs. Washington, DC, March 2000.

Shell Nederland BV
Ir. Henk J. Van Wouw, Manager Environmental Affairs. The Hague, November 2000.

Ministry of Housing, Spatial Planning and the Environment (VROM)
Barend van Engelenburg, Directorate-General for Environmental Protection, Directorate Climate Change and Industry. The Hague, November 2000.

European Commission
Marianne Wenning, Deputy Head, Climate Change Unit. Brussels, November 2000.

Global Climate Coalition (GCC)
Glenn F. Kelly, Executive Director and CEO. Washington, DC, March 2000.
Eric Hold, Washington, DC, March 2000.

American Petroleum Institute (API)
Phillip A. Cooney, Climate Team Leader. Washington, DC, March 2000.
William O'Keefe, Solutions Consulting. Washington, DC, March 2000.

European Petroleum Industry Association (EUROPIA)
Valèrie Callaud, Deputy Secretary General. Brussels, November 2000.

Pew Center on Global Climate Change
Eileen Claussen, Executive Director. Washington, DC, March 2000.
Sally C. Ericsson, Director of Outreach. Washinton, DC, March 2000.

Greenpeace International
Paul V. Horsman, Oil Campaigner, Greenpeace International Climate Campaign. Amsterdam, November 2000.

Greenpeace, US
Iain MacGill, Senior Policy Analyst, Greenpeace Climate Campaign. Washington, DC, March 2000.

World Resources Institute (WRI)
Jennifer Finlay, Director, Business Engagement, Management Institute for Environment and Business. Washington, DC, March 2000.
James MacKenzie, Senior Associate, Climate, Energy, and Pollution Program. Washington, DC, March 2000.
Kevin A. Baumert, Associate. Washington, DC, March 2000.

References

Aardal, B. and Valen, H. (1995), *Konflikt og opinion*, Oslo, NKS-forlaget.

Agrawala, S. (1998a), 'Context and early origins of the Intergovernmental Panel on Climate Change', *Climatic Change*, 39: 4, 605–20.

Agrawala, S. (1998b), 'Structural and process history of the Intergovernmental Panel on Climate Change', *Climatic Change*, 39: 4, 621–42.

Agrawala, S. and Andresen, S. (1999), 'Indispensability and indefensibility? The United States in climate treaty negotiations', *Global Governance*, 5, 457–82.

Agrawala, S. and Andresen, S. (2001), 'US climate policy: evolution and future prospects', *Energy and Environment*, 12: 2 and 3, 117–37.

Andersen, M. S. (1993), *Ecological Modernisation Between Policy Styles and Policy Instruments: The Case of Water Pollution Control*, Aarhus, Institute of Political Science, University of Aarhus.

Andersen, S. S. and Austvik, O. G. (2000), 'Petroleum og norske maktforhold: petroleum, nasjonal handlefrihet – nye internasjonale rammebetingelser', Report from ARENA-program, University of Oslo and Lillehammer College.

Andresen, S. and Hals Butenschøn, S. (2001), 'Norwegian climate policy: from pusher to laggard?', *International Environmental Agreements: Politics, Law and Economics*, 1, 337–57.

Andresen, S., Boehmer-Christiansen, S., Hanf, K., Kux, S., Lewanski, R., Morata, F., Skea, J., Sprinz, D., Underdal, A., Vaahtoranta, T. and Wettestad, J. (1996), *The Domestic Basis of International Environmental Agreements: Modelling National/International Linkages*, Final Report to the European Commission, Rotterdam, EC Contract EVSV-CT92-0185.

Arents, M. (1999), 'Dutch style of climate policy', Report, Twente, Centre for Clean Technology and Environmental Policy, University of Twente.

Arts, B. (2000), 'Regimes, non-state actors and the state system: a "structural" regime model', *European Journal of International Relations*, 6: 4, 513–42.

Baron, R. (1996), 'Policies and measures for common action', Working Paper on Economic/Fiscal Instruments, Paris, International Energy Agency.

Bartels, G. C. (1995), 'Main results of environmental communication studies', The Hague, Ministry of Housing, Spatial Planning and the Environment.

Benedick, R E. (1991), *Ozone Diplomacy – New Directions in Safeguarding Our Planet*, Cambridge, MA, Harvard University Press.

Bennett, G. (1991), 'Dutch policy plan', *Environment*, 33: 7, 7–9 and 31–3.

Bergesen, H. O., Roland, K. and Sydnes, A. K. (1995), *Norge i det globale drivhuset*, Oslo, Universitetsforlaget.

Betsill, M. M. and Corell, E. (2001), 'NGO influence in international environmental negotiations: a framework for analysis', *Global Environmental Politics*, 1: 4, 65–86.

Bolstad, G. (1993), *Inn i Drivhuset. Hva er Galt med Norsk Miljøpolitikk?*, Oslo, Cappelen.

Brunner, R. D. and Klein, R. (1998), *Harvesting Experience: A Reappraisal of the U.S. Climate Change Action Plan*, Boulder, Center for Public Policy Reasearch, University of Colorado.

Carpenter, C. (2001), 'Business, green groups and the media: the role of non-governmental organizations in the climate change debate', *International Affairs*, 77: 2, 313–28.

Christiansen, C. A. (2000), *On the Effectivness of Environmental Taxes: The Impacts of CO_2 Taxes on Environmental Innovation in the Norwegian Petroleum Sector*, FNI Report 10/2000, Lysaker, Fridtjof Nansen Institute.

Christiansen, C. A. (2002), 'New renewable energy developments and the climate change issue: a case study of Norwegian politics', *Energy Policy*, 30, 235–43.

Christiansen, C. A. and Wettestad J. (forthcoming), 'The EU as a frontrunner in greenhouse emissions trading: how did it happen and will the EU succeed?', Working Paper, Lysaker, Fridtjof Nansen Institute.

Corell, E. and Betsill, M. M. (2001), 'A comparative look at NGO influence in international environmental negotiations: desertification and climate change', *Global Environmental Politics*, 1: 4, 86–108.

Cox, R. W. and Jacobson, H. K. (1973), *The Anatomy of Influence: Decision Making in International Organization*, New Haven, CT, Yale University Press.

Dannenmaier, E. and Cohen, I. (2000), *Promoting Meaningful Compliance with Climate Change Commitments*, Washington, DC, Pew Centre.

Doremus, P. N., Keller, W. W., Pauly, L. W. and Reich, S. (1998), *The Myth of the Global Corporation*, Princeton, NJ, Princeton University Press.

Downs, A. (1972), 'Up and down with ecology – the issue attention cycle', *Public Interest*, 28: Summer, 38–50.

Dragsund, E., Aunan, K., Godal, O., Haugom, G. P. and Holtsmark, B. (1999), *Utslipp til Luft fra Oljeindustrien: Tiltak, Kostnader og Virkemidler*, Oslo, University of Oslo, CICERO-Report: 2.

DSP (Department of State Publications) (1997), *Climate Action Report. 1997 Submission of the United States of America Under the UNFCCC*, Washington, DC, Department of State Publications 10496, Bureau of Oceans and International Environmental Scientific Affairs, Office of Global Change.

Dunlap, R. E. (2000), *Americans have Positive Image of the Environmental Movement*, Poll releases, 18 April 2000, source: www.gallup.com/poll/releases/pr000418.asp (accessed 11 May 2000).

ECCP (European Climate Change Programme) (2001), 'European Climate Change Programme, report – June 2001', source: http://europa.eu.int/comm/environment/climat/home_en.htm (accessed 25 February 2002).

Eikeland, P. O. (1993), 'US energy policy at the cross-roads?', *Energy Policy*, 21: 10, 987–98.

Eikeland, P. O. (forthcoming), 'Environmental innovation in the Norwegian and Swedish electricity supply industry – the role of "national energy-industrial styles"', Report, Sandvika, Norwegian School of Management.

EPA (United States Environmental Protection Agency) (2000), *The U.S. Greenhouse Gas Inventory*, Washington, DC, EPA Office of Air and Radiation.

Estrada, J., Tangen, K. and Bergesen, H. O. (1997), *Environmental Challenges Confronting the Oil Industry*, Chichester, John Wiley and Sons.

EU Commission, COM (1997) 599 final: Communication from the Commission, 'Energy for the future: renewable sources of energy', White Paper for a Community strategy and action plan (26 November 1997).

EU Commission, COM (2000a) 749: 'Report under the Council Decision

1999/296/EC for a monitoring mechanism of Community greenhouse gas emissions'.

EU Commission, COM (2000b) 87 final: 'Green Paper on greenhouse gas emissions trading within the European Union (presented by the Commission)', Brussels (8 March 2000).

EU Commission, COM (2001) 581 final: 'Proposal for a Directive of the European Parliament and the Council establishing a scheme for greenhouse gas emission allowance trading within the Community and amending Council Directive 96/61/EC', Brussels (23 October 2001).

EUROPIA (1999), *Activity Report*, Brussels, EUROPIA.

Evans, P. B, Jacobson, H. K. and Putnam, R. D. (eds.) (1993), *Double-Edged Diplomacy: International Bargaining and Domestic Politics*, Berkeley, University of California Press.

ExxonMobil (1998), *Annual Report*, source: www.exxonmobil.com.

ExxonMobil (1999a), *Annual Report*, source: www.exxonmobil.com.

ExxonMobil (1999b), *Financial and Operating Review*, source: www.exxonmobil.com.

ExxonMobil (2000a), *Annual Report*, source: www.exxonmobil.com.

ExxonMobil (2000b), *Financial and Operating Review, 2000*, source: www.exxonmobil.com.

ExxonMobil (2000c), *ExxonMobil Safety, Health and Environment Progress Report 2000: The People Behind the Commitment*, source: www.exxon.mobil.com/news/publications/c_she/c_page2.html.

ExxonMobil (2001a), *ExxonMobil Safety, Health and Environment Progress Report 2001*, source: www.exxonmobil.com.

ExxonMobil (2001b), *Global Climate Change: ExxonMobil Views*, April 2001, source: www.exxon.mobil.com.

Falkner, R. (1996), 'The roles of firms in international environmental politics', paper presented at the 37th Annual Convention of the International Studies Association in San Diego, California, USA, 16–20 April. Oxford, Nuffield College.

Farsund, A. A. (1997), 'Den globale utfordring' ('The global challenge') in J. E. Klausen and H. Rommetvedt (eds.), *Miljøpolitikk: Organisasjonene, Stortinget og forvaltningen*, Oslo, Tano Aschehoug.

Fay, C. (1997), 'Achievable targets needed', speech by Chris Fay (Chairman and Chief Executive Shell U.K. Limited) at the CBI Panel Debate, 'Climate change – a taxing business?' at the CBI National Conference, Birmingham, 10 November 1997, source: www.shell.co.uk/news/speech/spe_achievable.htm (accessed 22 November 1999).

Flannery, B. P. (1999), 'Global climate change', *International Association for Energy Economics*, Newsletter, Third Quarter, 4–10.

Gallup, A. and Saad, L. (1997), 'Public concerned, not alarmed about global warming', Poll Releases, 2 December 1997, source: www. gallup.com/poll/releases/pr971202.asp (accessed 10 May 2000).

Ghobadian A., Viney, H., Liu, J. and James, P. (1998), 'Extending linear approaches to mapping corporate environmental behaviour', *Business Strategy and the Environment*, 7, 13–23.

Gjerde, K. (1992), 'Norsk foruresningspolitikk – et uttrykk for en "avveiningstankegang". En studie av SFT's regulering av stor-forurensende industri', Hovedoppgave i Administrasjon og organisasjonsvitenskap, Bergen, University of Bergen.

Gleckman, H. (1995), 'Transnational corporations' strategic responses to "sustainable development"', *Green Globe Yearbook*, Oxford, Fridtjof Nansen Institute and Oxford University Press.

Grant, W., Matthews, D. and Newell, P. (2000), *The Effectiveness of European Union Environmental Policy*, London, Macmillian.

Greenpeace International (1998), 'The oil industry and climate change', Greenpeace Briefing by Kirsty Hamilton.

Greenpeace International (2001), 'Greenpeace to target US Oil companies', press release, 26 April.

Grolin, J. (1998), 'Corporate legitimacy in risk society: the case of Brent Spar', *Business Strategy and the Environment*, 7, 213–22.

Grubb, M., Vrolijk, C. and Brack, D. (1999), *The Kyoto Protocol: A Guide and Assessment*, London, Royal Institute of International Affairs.

Haas, P. M. (1991), 'Policy responses to stratospheric ozone depletion', *Global Environmental Change*, June, 224–34.

Hass, J. L. (1996), 'Environmental ("green") management typologies: an evaluation, operationalization and empirical development', *Business Strategy and the Environment*, 5, 59–68.

Hatch, M. T. (1993), 'Domestic politics and international negotiations: the politics of global warming in the United States', *Journal of Environment and Development*, 2: 29, 1–39.

Haufler, V. (1993), 'Crossing the boundary between public and private: international regimes and non-state actors', in V. Rittberger and P. Mayer (eds.), *Regime Theory and International Relations*, Oxford, Clarendon Press.

IEA (International Energy Agency) (1994), *Climate Change Policy Initiatives: Volume I OECD Countries*, Paris, International Energy Agency.

IEA (International Energy Agency) (1997), *Voluntary Actions for Energy-Related CO_2 Abatement*, Paris, International Energy Agency/Organization for Economic Co-operation and Development.

Ikwue, T. and Skea, J. (1994), 'Business and the genesis of the European Community carbon tax proposal', *Business Strategy and the Environment*, 3: 2, 1–11.

Inglehart, R. (1971), 'The silent revolution in Europe: intergenerational change in post-industrial societies', *American Political Science Review*, 65, 991–1017.

Jänicke, M. (1992), 'Conditions for environmental policy success: an international comparison', in M. Jachtenfuchs and M. S. Strübel (eds.), *Environmental Policy in Europe: Assessment, Challenges and Perspectives*, Baden-Baden, Nomos.

Jänicke, M. (1997), 'The political system's capacity for environmental policy', in M. Jänicke and H. Weidner (eds. in collaboration with H. Jörgens), *National Environmental Policies: A Comparative Study of Capacity-Building*, Berlin, Springer.

Jansen, A. I. and Osland, O. (1996), 'Norway', in *Governing the Environment: Politics, Policy, and Organization in the Nordic Countries*, Nord: 5, Copenhagen, Nordic Council of Ministers.

Kasa, S. (1999), 'Social and political barriers to green tax reform: the case of CO_2 taxes in Norway', CICERO Policy Note no. 5, Oslo, University of Oslo.

Kasa, S. (2000), 'Policy networks as barriers to green tax reform: the case of CO_2 taxes in Norway', *Environmental Politics*, 9: 4, 104–22.

Kasa, S. and Malvik, H. (2000), 'Makt, miljøpolitikk, organiserte industriinteresser og partistrategier: en analyse av de politiske barrierene mot en utvidelse av CO_2-avgiften i Norge' ('Power, environmental politics, industry interests and party strategies: an analysis of the political barriers to an expansion of the CO_2 tax in Norway'), *Tidsskrift for Samfunnsforskning*, 3, 295–323.

Ketola, T. (1993), 'The seven sisters: Snow Whites, dwarfs or evil queens? A comparison of the official environmental policies of the largest oil corporations in the World', *Business Strategy and the Environment*, 2: 3, 22–33.

Kirby, E. G. (1995), 'An evaluation of the effectiveness of US CAFÉ policy', *Energy Policy*, 23: 2, 107–9.

Kolk, A. (2001), 'Multinational enterprises and international climate policy', in B. Arts, M. Noortmann and B. Reinalda (eds.), *Non-State Actors in International Relations*, Aldershot, Ashgate.

Kolk, A. and Levy, D. (2001), 'Winds of change: corporate strategy, climate change and oil multinationals', *European Management Journal*, 19: 5, 501–9.

Koplow, D. and Martin, A. (1999), 'Fueling global warming: federal subsidies to oil in the United States', Washington, DC, Greenpeace Report.

Kotvis, J. A. (1994), 'The economics of environmental issues in international oil and gas exploration and production', *Petroleum Accounting and Financial Journal*, 13: 3, 88–97.

Leggett, J. (1999), *The Carbon War: Dispatches from the End of the Oil Century*, London, Penguin.

Levy, D. (1997), 'Business and international environmental treaties: ozone depletion and climate change', *California Management Review*, 39: 3, 54–71.

Levy, D. L. and Egan, D. (1998), 'Capital contests: national and transnational channels of corporate influence on the climate change negotiations', *Politics and Society*, 26: 3, 337–61.

Levy, D. L. and Newell, P. (2000), 'Oceans apart: business responses to global environmental issues in Europe and the United States', *Environment*, 42: 9, 8–21.

Levy, D. L. and Rothenberg, S. (1999), 'Corporate strategy and climate change: heterogeneity and change in the global automobile industry', ENRP Discussion Paper E-99-13, Boston, Kennedy School of Government, Harvard University.

Levy, M. A. (1993), 'European acid rain: the power of tote-board diplomacy', in P. M. Haas, R. O. Keohane and M. A. Levy (eds.), *Institutions for the Earth: Sources of Effective International Environmental Protection*, Cambridge, MA, MIT Press.

Levy, M. A., Young, O. R. and Zürn, M. (1995), 'The study of international regimes', *European Journal of International Relations*, 1: 3, 267–330.

Liefferink, D. (1995), 'Environmental policy in the Netherlands: national profile', paper prepared for the workshop 'New Nordic member states and the impact on EC environmental policy', Sandbjerg, Denmark, 6–8 April, Wageningen Agricultural University. Later published as Liefferink, D. (1997), 'The Netherlands: a net exporter of environmental policy concepts', in M. S. Anderson and D. Liefferink (eds.), *European environmental policy: the pioneers*. Manchester, Manchester University Press, pp. 210–50.

Loreti, C. P., Foster, S. A., Obbagy, J. E. and Little, A. D. (2001), 'An overview of greenhouse gas emissions verification issues', Washington, DC, Pew Centre.

Lucardie, P. (1999), 'Dutch politics in the late 1990s: "purple" government and "green" opposition', *Environmental Politics*, 8: 3, 153–8.

Lundqvist, L. J. (1980), *The Hare and the Tortoise: Clean Air Policies in the United States and Sweden*, Ann Arbor, MI, University of Michigan Press.

Mahlman, J. D. (1997), 'Uncertainties in projections of human-caused climate warming', *Science*, 278, 21 November, 1416–17.

Maloney, W. A., Jordan, G. and McLaughlin, A. M. (1994), 'Interest groups and public policy: the insider/outsider model revisited', *Journal of Public Policy*, 14: 1, 17–38.

Marriott, P. (1991), 'Energy planning short term (to 2010)', in B. P. Flannery and R. Clarke (eds.), *Global Climate Change: A Petroleum Industry Perspective*, London, IPIECA (International Petroleum Industry Environmental Conservation Association).

Maxwell, J. H. and Weiner, S. L. (1993), 'Green consciousness or dollar diplomacy? The British response to the threat of ozone depletion', *International Environmental Affairs*, 5: 1, 19–41.

Menz, F. C. (2002), 'Bush proposal on climate change: business as usual?', *Cicerone* newsletter 1/2002, source: www.cicero.uio.no (accessed 25 September 2002).

Miles, E. D., Underdal, A., Andresen, S., Wettestad, J., Skjærseth, J. B. and Carlin, E. M. (2001), *Environmental Regime Effectiveness: Confronting Theory with Evidence*, Cambridge, MA, MIT Press.

MILJØSOK (1996), 'Oljeindustrien tar ansvar, rapport fra Styringsgruppen' ('The oil industry takes responsibility, report from the steering group'), Oslo, MILJØSOK.

MILJØSOK (2000), 'Felles miljø, felles satsninger' ('Common environment, common priorities'), Oslo, MILJØSOK.

Ministry of Environment (1997), 'Norway's second national communication under the UNFCCC', Oslo, Ministry of Environment.

Neale, A. (1997), 'Organisational learning in contested environments: lessons from Brent Spar', *Business Strategy and the Environment*, 6, 93–103.

Newell, P. N. (2000) *Climate for Change: Non-state Actors and the Global Politics of the Greenhouse*, Cambridge, Cambridge University Press.

Nogepa (1998), 'Voorgangsrapportage ontwikkeling enrgie-efficiency', The Hague, Nederlandse olie- en gaswinnings-industrie.

Noreng, Ø. (1996), 'National oil companies and their government owners: the politics of interaction and control', *Journal of Energy and Development*, 19: 2, 197–226.

Norwegian White Paper no. 46 (Stortingsmelding) (1988–1989), 'Miljø og utvikling. Norges oppfølging av Verdenskommisjonens repport', Oslo, Ministry of Environment.

Norwegian White Paper no. 41 (Stortingsmelding) (1994–1995), 'Om norsk politikk mot klimaendringer og utslipp av nitrogenoksider (NO_x)', Oslo, Ministry of Environment.

Norwegian White Paper no. 29 (Stortingsmelding) (1997–1998), 'Norges oppfølging av Kyotoprotokollen', Oslo, Ministry of Environment.

Norwegian White Paper no. 29 (Stortingsmelding) (1998–1999), 'Om energipolitikken', Oslo, Ministry of Petroleum and Energy.

Novem (Netherlands Agency for Energy and the Environment) (1998), 'Voortgangsrapportage: energie-efficiency in de Nederlandse aardolieindustrie', The Hague, Netherlands Agency for Energy and the Environment.

Nuijen, W. C. (1999), 'Experience with long term agreements on energy efficiency and outlook to policy for the next decade', paper presented at the Energy 2000 Conference, 27 August 1999, Basle, Switzerland. The Hague, Netherlands Agency for Energy and the Environment (Novem).

OECD (Organisation for Economic Co-operation and Development) (1996), *Environmental Performance Reviews, United States*, Paris: Organisation for Economic Co-operation and Development.

OECD (Organisation for Economic Co-operation and Development) (1999), *Voluntary Approaches for Environmental Policy*, Paris: Organisation for Economic Co-operation.

Paterson, M. (1996), *Global Warming and Global Politics*, London: Routledge.

Paterson, M. (1999), 'Global finance and environmental politics: the insurance industry and climate change', *IDS Bulletin*, 30: 3, 25–30.

Pauly, L. W. and Reich, S. (1997), 'National structures and multinational corporate behaviour: enduring differences in the age of globalization', *International Organization*, 51: 1, 1–30.

Perkins, J. M. (1999), 'Economic state of the US petroleum industry', Washington, DC, American Petroleum Institute.

Post, J. E. and Altman, B. W. (1992), 'Models of corporate greening: how corporate social policy and organizational learning inform leading-edge environmental management', *Research in Corporate Social Performance and Policy*, 13, 3–29.

Powell, W. W. and DiMaggio, P. J. (eds.) (1991), *The New Institutionalism in Organizational Analysis*, Chicago and London, University of Chicago Press.

Putnam, R. D. (1988), 'Diplomacy and domestic politics: the logic of two-level games', *International Organization*, 42: 3, Summer, 427–60.

Ragin, C. (1987), *The Comparative Method*, Berkeley, University of California Press.

Raustiala, K. (2001), 'Nonstate actors in the global climate regime', in U. Lutherbacher and D. F. Sprinz (eds.), *International Relations and Global Climate Change*, Cambridge, MA, MIT Press.

Reitan, M. (1998), 'Interesser og institusjoner i miljøpolitikken' ('interests and institutions in environmental policy'), PhD disseratation 5/98, Oslo, Akademika and the University of Oslo.

Retallack, S. (2000), 'Economic globalization and the environment', *Transnational Association*, 4, 181–91.

Ringius, L. (1997), *Differentiation, Leaders and Fairness: Negotiating*

Climate Commitments in the European Community, CICERO Report no. 8, Oslo, University of Oslo.

Risse-Kappen, T. (1995), *Bringing Transnational Relations Back In: Non-State Actors, Domestic Structures and International Institutions*, Cambridge, Cambridge University Press.

Rokkan, S. (1966), 'Norway: numerical democracy and corporate pluralism', in R. A. Dahl (ed.), *Political Opposition in Western Democracies*, New Haven, CT, Yale University Press.

Rondinelli, D. A. and Berry, M. A. (2000), 'Environmental citizenship in multinational corporations: social responsibility and sustainable development', *European Management Journal*, 18: 1, 70–84.

Roome, N. (1992), 'Developing environmental strategies', *Business Strategy and the Environment*, 1: 1, 11–25.

Rowlands, I. H. (2000), 'Beauty and the beast? BP's and Exxon's positions on global climate change', *Environment and Planning C: Government and Policy*, 18: 3, 339–54.

Rudsar, K. (1999), 'Liberalisering av norsk petroleumspolitikk: et resultat av samspillet mellom internasjonale og nasjonale faktorer?', *Internasjonal Politikk*, 57: 3, 429–50.

Saad, L. and Dunlap, R. E. (2000), 'Americans are environmentally friendly, but issue not seen as urgent problem', Poll Releases 17 April 2000, source: www.gallup.com/poll/releases/pr000417 (accessed 10 May 2000).

Sabatier, P. (1986), 'Top-down and bottom-up approaches to implementation research: a critical analysis and suggested synthesis', *Journal of Public Policy*, 6: 1, 21–48.

Sand, P. H. (1991), 'Lessons learned in global environmental governance', *Environmental Affairs Law Review*, 18, 213–77.

Schelling, T. C. (1978), *Micromotives and Macrobehaviour*, New York and London, W. W. Norton.

Schenkel, W. (1998), *From Clean Air to Climate Policy in the Netherlands and Switzerland – Same Problems, Different Strategies?*, Berlin, Peter Lang.

Seippel, Ø. and Lafferty, W. M. (1996), 'Religion, menneske eller økologi: om natursyn og miljøpolitisk engasjement', *Norsk Statsvitenskapelig Tidsskrift*, 12: 2, 113–39.

Shell (1996), 'Annual report', source: www.shell.com.

Shell (1998a), 'Annual report', source: www.shell.com.

Shell (1998b), 'The 1998 Shell report', source: www.shell.com.

Shell (1998c), 'Shell, global scenarios 1998–2020', source: www.shell.com.

Shell (1999), 'Annual report', source: www.shell.com.

Shell (2000a), 'The 2000 Shell report', source: www.shell.com.

Shell (2000b), 'Annual Report', source: www.shell.com.

Shell (2000c), 'Shell Tradeable Emission Permit System (STEPS)', source: www.shell.com.

Shell (2001), 'The 2001 Shell Report', source: www.shell.com.

Skea, J. (1992), 'Environmental issues facing the oil industry', *Energy Policy*, 20: 10, 950–8.

Skjærseth, J. B. (1992), 'The "successful" ozone layer negotiations: are there any lessons to be learned?', *Global Environmental Change*, December, 292–300.

Skjærseth, J. B. (1994), 'The climate policy of the EU: too hot to handle?', *Journal of Common Market Studies*, 32: 1, March, 25–45.

Skjærseth, J. B. (1995), 'The fruitfulness of various models in the study of international environmental politics', *Cooperation and Conflict*, 30: 2, 155–78.

Skjærseth, J. B. (1999), *The Making and Implementation of North Sea Pollution Commitments: Institutions, Rationality and Norms*, Oslo, Department of Political Science, University of Oslo and Akademika AS.

Skjærseth, J. B. (2000), 'Environmental "voluntary" agreements: conditions for making them work', *Swiss Political Science Review*, 6: 2, 57–78.

Skjærseth, J. B. and Rosendal, K. (1995), 'Norges miljø-utenrikspolitikk' ('Norway's environmental foreign policy'), in T. L. Knutsen, G.M. Sørbø and S. Gjerdåker (eds.), *Norges utenrikspolitikk*, Oslo, Cappelens Akademisk forlag.

Skjærseth, J. B. and Skodvin, T. (2001), 'Climate change and the oil industry: common problems, different strategies', *Global Environmental Politics*, 1: 4, 43–64.

Skjærseth, J. B. and Wettestad, J. (2002), 'Understanding the effectiveness of EU environmental policy: how can regime analysis contribute?', *Environmental Politics*, 11: 3, 99–120.

Skjåk, K. K. and Bøyum, B. (1996), 'Haldningar til offentlege styringsmakter og offentlig verksemd', Report no. 108, Bergen, Norsk samfunnsvitenskapelig datatjeneste (NSD).

Skodvin, T. (2000a), *Structure and Agent in Scientific Diplomacy*, Dordrecht, Kluwer Academic.

Skodvin, T. (2000b), 'The Intergovernmental Panel on Climate Change', in S. Andresen, T. Skodvin, J. Wettestad and A. Underdal, *Science and Politics in International Environmental Regime: Between Integrity and Involvement*, Manchester, Manchester University Press.

Skodvin, T. and Skjærseth, J. B. (2001), 'Shell Houston, we have a climate problem!', *Global Environmental Change*, 11: 2, 103–6.

Skolnikoff, E. B. (1997), 'Same science, differing policies: the saga of global climate change. MIT joint program on the science and policy of global change', Report no. 22, Cambridge, MA, MIT Press.

Slingerland, S. (1997), 'Energy conservation and organisation of electricity supply in the Netherlands', *Energy Policy*, 25: 2, 193–203.

Statoil (1992), 'Statoils miljørapport for 1991', Stavanger: Den Norske Stats Oljeselskap.

Statoil (1997), 'Statoil 1997 health, safety and environment report', source: www.statoil.com.

Statoil (1998), 'Annual report 1998', source: www.statoil.com.

Statoil (1999a), 'Annual report 1999', source: www.statoil.com.

Statoil (1999b), 'Creating value for Statoil and the SDFI', report presented to the minister of petroleum and energy by Statoil's board of directors, 13 August 1999, source: www.statoil.com.

Statoil (2000), 'Annual report 2000', source: www.statoil.com.

Statoil (2001), 'Annual report 2001', source: www.statoil.com.

Steger, U. (1993), 'The greening of the board room: how German companies are dealing with environmental issues', in K. Fischer and J. Schot (eds.), *Environmental Strategies for Industry: International Perspectives on Research Needs and Policy Implications*, Washington, DC, Island Press.

Suurland, J. (1994), 'Voluntary agreements with industry: the case of Dutch covenants', *European Environment*, 4: 4, 3–7.

Sydnes, A. K. (1996), 'Norwegian climate policy: environmental idealism and economic realism', in T. O'Riordan and J. Jäger (eds.), *Politics of Climate Change: A European Perspective*, London, Routledge.

Tak, T. (1994), 'Shades of green: political parties and Dutch environmental policy', in M. Wintle and R. Reeve (eds.), *Rhetoric and Reality in Environmental Policy: The Case of the Netherlands in Comparison with Britain*, Aldershot, Avebury.

Tenfjord, A. P. (1995), *CO₂-saka: Økonomisk Interessepolitikk og Miljøpolitiske Målsetjingar* (*The CO₂ Case: Economic Interest Policy and Environmental Objectives*), Report 9507, Bergen, LOS-senteret.

Underdal, A. (1980), *The Politics of International Fisheries Management: The Case of the Northeast Atlantic*, Oslo, Universitetsforlaget.

Underdal, A. (1984), 'Can we, in the study of international politics, do without the model of the state as a rational, unitary actor? A discussion of the limitations and possible fruitfulness of the model, and its alternatives', *Internasjonal Politikk*, Temahefte I, NUPI, 63–79.

Underdal, A. (1992), 'The concept of regime "effectiveness"', *Cooperation and Conflict*, 27: 3, 227–40.

Underdal, A. (1998), 'Explaining compliance and defection: three models', *European Journal of International Relations*, 4: 1, 5–30.

Underdal, A. (2001), 'One question, two answers', in E. D. Miles, A. Underdal, S. Andresen, J. Wettestad, J. B. Skjærseth and E. M. Carlin, *Environmental Regime Effectiveness: Confronting Theory with Evidence*, Cambridge, MA, MIT Press.

Underdal, A, and Hanf, K. (eds.) (2000), *International Environmental Agreements and Domestic Politics: The Case of Acid Rain*, Aldershot, Ashgate.

UNICE (1991), 'Position on a number of basic principles for the formulations of a Community action strategy on the greenhouse effect', 25 June, Brussels, UNICE.

UNICE (2000), 'UNICE input ahead of COP 6 at The Hague', November, Brussels, UNICE.

UNICE (2002), 'UNICE comments on the proposal for a framework for EU emissions trading', Brussels: UNICE.

Vedung, E. (1997), *Policy Instruments: Typologies and Theories*, Uppsala, Uppsala University, Department of Government.

Vogel, D. (1986), *National Styles of Regulation: Environmental Policy in Great Britain and the United States*, London, Cornell University Press.

VROM (1989), *National Environmental Policy Plan (NEPP)*, The Hague, Ministry of Housing, Physical Planning and Environment.

VROM (1991), *National Environmental Policy Plan Plus*, The Hague, Ministry of Housing, Physical Planning and Environment.

VROM (1993), *Second National Environmental Policy Plan (NEPP, 2)*, The Hague, Ministry of Housing, Physical Planning and Environment.

VROM (1997), *Netherlands' National Communication on Climate Change Policies (second), 1997*, The Hague, Ministry of Housing, Physical Planning and Environment.

VROM (1998), *Third National Environmental Policy Plan (NEPP, 3)*, The Hague: Ministry of Housing, Physical Planning and Environment.

VROM (1999), *The Netherlands' Climate Policy Implementation Plan, Part I: Measures in the Netherlands*, The Hague, Ministry of Housing, Spatial Planning and the Environment.

Wallace, D. (1995), *Environmental Policy and Industrial Innovation: Strategies in Europe, the US and Japan*, London, Earthscan.

Waltz, K. N. (1979), *Theory of International Politics*, Reading, MA, Addison-Wesley.

Watts, P. (1997), 'Taking action to earn trust – health, safety and the environment', speech, issued 22 September 1997, source: www.shell.com/library/speech/0,1525,2304,00.html.

Weale, A. (1992), *The New Politics of Pollution*, Manchester, Mancester University Press.

Werf, T. (ed.) (2000), 'Climate change: solution in sight. A Dutch perspective', Delft, Dutch Energy Policy Platform.

Wettestad, J. (2001), 'The ambiguous prospects for EU climate policy – a summary of options', *Energy and Environment*, 12: 2 and 3, 139–67.

Yin, R. K. (1989), *Case Study Research: Design and Methods*, London, SAGE.

Young, O. R. (1989), *International Cooperation: Building Regimes for Natural Resources and the Environment*, Ithaca, NY, Cornell University Press.

Young, O. R. (1994), *International Governance: Protecting the Environment in a Stateless Society*, Ithaca, NY, and London, Cornell University Press.

Young, O. R. and Osherenko, G. (1993), *Polar Politics: Creating International Environmental Regimes*, Ithaca, NY, Cornell University Press.

Index

Page references for tables and figures are in *italics*.

Aardal, B. 26, 113
Ad Hoc Group on the Berlin
    Mandate (AGBM) 169
Agrawala, S. 110, 118, 139, 162,
    164, 165, 166, 170, 173, 177
ALTERNER programme 184
Altman, B. W. 3, 20
American Energy Alliance 139
American Petroleum Institute
    (API) 49, 52, 137, 172, 216
    BTU tax 139, 140
    Clinton–Gore administration
        138, 172–3
    Kyoto Protocol 165, 166
    UNFCCC 161
Andresen, S. 37, 110, 118, 139,
    162, 164, 165, 166, 170, 173
Arctic Ocean 81, 82
authoritative force 27–8
auto industry 212

Barland, Knut 63
Bellona 82
benchmark agreements (BAs) 127
Bennett, G. 122, 141
Bergesen, H. O. 118, 129
Berlin Mandate 164, 169, 217
Berry, M. A. 3, 25, 28
best available technology (BAT)
    147

Betsill, M. M. 35, 163, 168, 169,
    172
biofuels 64
Biovarme 64
Boren, Senator 139
BP 10
    capital availability 98
    climate strategies 1, 4–5, 55, 60
    emissions trading 183, 187
    environmental reputation 81, 82
    GCC 167, 175, 210, 216
    GHG emissions 57
    human resource availability 99
    leadership 23, 97
    ownership 99
    policy instruments 134
    renewables 185
Brent Spar 80, 105, 150, 208
British thermal unit (BTU) tax
    118, 133–4, 139, 140, 150,
    162, 164, 215
Browne, Sir John 23, 97
Brundtland, Gro Harlem 128,
    145–6
Buenos Aires Plan of Action 181,
    217–18
Bush, George W. 165, 180, 214,
    218
Bush–Cheney administration 120,
    139, 140, 182, 211

Bush–Quayle administration
    117–18, 163
business reorientation 15, 43, 66,
    68–9, 69, 197, 200–1
    ExxonMobil 51, 52
    Shell 57–9
    Statoil 64–5
Byrd/Hagel resolution 165,
    167–8, 173, 175, 190, 203–4,
    215

CA model *see* Corporate Actor
    model
Canada 218
capabilities 35
capital availability 22, 23, 98,
    198
carbon dioxide emissions
    auto industry 212
    GCC 164
    Netherlands 122, 123, 124,
        127, 141, 184
    Norway 128, 129, 131, 132–3
    Statoil 63–4, 65, 67
    UK 134
carbon intensity 76–7, 78, 95, 96
carbon storage 51
Carpenter, C. 164, 168, 169
causal complexity 8–9, 22, 207
CFCs 6
change 151, 152, 153, 158–9,
    168–72, 188–9, 203
chemical industry 6, 33
Christiansen, C. A. 130, 131,
    133, 146, 147, 183
Clean Air Act (US) 136–7
Clean Development Mechanism
    (CDM) 56, 168–9
Climate Action Network (CAN)
    169
climate change 15, 43, 66, 69,
    158, 197, 200
    ExxonMobil 46, 48, 49–50, 51
    media coverage 110
    Shell 54, 59
    Statoil 62, 65
Climate Change Action Plan
    (CCAP) 118–19, 120

Climate Change Programme (UK)
    134
Climate Network Europe (CNE)
    169, 171–2
climate policy 104, 116–17,
    133–5, 212–18
    Netherlands 121–5
    Norway 128–32, 153
    US 117–20, 151, 152
climate strategies 1–2, 12–13,
    40–1, 43–4, 69, 196,
    197–200
    alternative approaches 209–11
    analytical implications 204–9
    comparison 66–70
    Corporate Actor model 18–23
    distinctions 13–16
    Domestic Politics model 24–31
    ExxonMobil 46, 48–52, 200–1
    International Regime model
        31–40
    policy implications 211–18
    research strategy 7–9
    Shell 54–60, 200–1
    Statoil 62–6, 200–1
Clinton–Gore administration
    BTU tax 150, 162, 164
    climate policy 118–20, 133–4,
        139, 213
    corporate influence 137–8, 140,
        152, 172–3, 175, 214, 215
    Kyoto Protocol 164–5, 166,
        190
    two-level games 211
CO_2 emissions *see* carbon dioxide
    emissions
coal 19
    ExxonMobil 76–7
    Shell 57, 58, 59, 76–7
command and control 27, 28
Committee on Environment and
    Industry (Netherlands) 141
Concern for Tomorrow 122
conflict-oriented approach 135,
    136–40, 149
consensual approach 135, 137–8,
    140–3, 149
consensus 36, 142

consumer behaviour 25, 26
Corell, E. 35, 163, 168, 169, 172
Corporate Actor model 8, 12, 18,
    41, 74–5, 158, 198, 199,
    204–5, 207
  company-specific factors 100–1,
    104, 201–2
  and DP model 207–8
  environmental reputation
    19–20, 78–82
  environmental risk 18–19, 75–8
  explanatory power 94–6, *94*
  moderating factors 22–3,
    96–100
  organisational learning 20–2,
    83–8, *89*, 90, *91*, 92, 93–4
corporate climate strategies *see*
    climate strategies
corporate influence 159–60,
    174–6, 188–91, 192–3,
    208–9, 214–15
  access and decision-making
    procedures 172–4
  changes 163–4
  counterbalancing forces 168–72
  fossil-fuel lobby 164–8
  and proactive strategies 215–16
  and reactive strategies 216–17
  UNFCCC 160–3
corporate leadership 22, 23, 97–8,
    198, 205
corporations 2–7, 16–18
corporative channel 29–31, 135
cost-benefit analyses 50
Council for the Environment
    (Netherlands) 140
courts 137
critical mass phenomena 210

decarbonisation 14, 57, 58, 68
decision-making 36–7, 173–4
  Shell 88, 142
defensive strategies *see* reactive
    strategies
Denmark 162
Domestic Politics (DP) model 8,
    12–13, 24, 41, 158, 198–200,
    202–3, 205–6

and CA model 207–8
context 18, 104–5, 148–53, *148*
governmental supply 26–9,
    116–35
interplay 207
and IR model 208–9
political institutions 29–31,
    135–47
social demand 24–6, 105–16
two-level games 211
Downs, A. 25, 26, 110, 114
Dunlap, R. E. *109*, 110, 111, 112

Eco-Management and Audit
    Scheme (EMAS) 54
economic instruments 27
Egan, D. 6, 16, 29, 110–11
emissions trading 40
  ENGOs 168–9, 171–2
  EU 183–4, 186–8, 192, 204
  Norway 130
  Shell *55–6*, 59
  UK 134
  see also greenhouse gas
    emissions
energy efficiency 14
  ExxonMobil 50
  Netherlands 125–7, *126*, 141,
    142–3
  Norway 132
  Shell 56, 68
energy taxes
  EU 162, 166, 171
  Netherlands 123, 124, 142
  Norway 128, 129, 130, 131,
    132–3, 135, 145–7, 149, 153,
    213
  US 118, 133–4, 139, 140, 150,
    162, 164, 215
environmental non-governmental
    organisations (ENGOs) 5, 37,
    112–13, 159, 163, 213, 219
  influence 168–72
  Netherlands 108
  Norway 114, 115, 153
  US 111–12
Environmental Protection Act
    (Netherlands) 140

Environmental Protection Agency
    (EPA) (US) 121, 136–7
environmental reputation 19–20,
    83, 198
  company-specific factors 74,
    78–82, 95, 96
environmental risk 18–19, 83,
    198
  company-specific factors 74,
    75–8, 95, 96, 201
Environmental Tax Committee
    (Norway) 145
Esso 80–1, 112, 113, 114
Estrada, J. 19, 28, 44, 45, 53, 61,
    85, 86, 109, 113
European Climate Change
    Programme (ECCP) 183, 186,
    192, 204
European Commission 171, 184,
    186, 187, 213
European Court of Justice 37
European Environment Bureau
    (EEB) 171, 213
European Parliament 174, 184
European Union (EU) 39, 204
  climate policy 124–5, 188
  corporate influence 160, 161–2,
    216–17
  decision-making 37, 173–4,
    175, 188–9
  EMAS 54
  emissions trading 186–8, 192
  energy taxes 143
  ENGOs 170–2
  Kyoto Protocol 166–7, 182–4,
    208
  leadership 218
  pressure 191–2
  renewables 184–6, 192
EUROPIA (European Petroleum
    Industry Association) 143,
    162, 184, 186, 187–8,
    216–17
explanation-building 9
Exxon 44, 88, 98, 210
Exxon Energy Cube 110–11
Exxon Valdez 19–20, 46, 79, 82,
    95, 110

ExxonMobil 1, 9, 10, 44–5, 158,
    196, 208, 215
  CA model 94–6, *94*, 101
  capital availability 98
  climate strategies 4–5, 43–4,
    46, 48–52, 66–70, *69*, 74,
    200–1
  corporate influence 216–17
  DP model 105, 148, 149,
    151–2, 153, 202–3
  emissions trading 187–8
  environmental policy 45–6, 47
  environmental reputation 79,
    80–1, 82
  environmental risk 75–8, *77*,
    201
  GCC 164
  GHG emissions 219
  human resource availability 99
  Kyoto Protocol 163, 165,
    167–8, 175–6, 203–4
  leadership 97
  monitoring of trends 85–6, 87
  organisational learning 93–4
  organisational structure 88, 90,
    91
  ownership 23, 99, 100
  policy instruments 120–1, 125,
    128, 134
  and political institutions 136,
    138, 139, 140
  reactive strategies 188, 197
  science 176, 180–1, 191
  social demand 105, 106,
    109–13, *109*, 114, 115–16
  and US government 119–20

Falkner, R. 6, 26
Fay, Chris 56, 57, 58–9
Ford Motor Company 167, 212
fossil-fuel lobby 164–8, 174–5
  see also corporate influence
Friends of the Earth (FoE) 82,
    171
fuel-cells 64

gas *see* natural gas
General Motors 167

Ghobadian, A. 3, 23
Gleckman, H. 2, 3, 6, 24
Global Climate Change
     Symposium 85
Global Climate Coalition (GCC)
     138–9, 210, 212, 216
   BP 175
   ExxonMobil 49, 52
   fossil-fuel lobby 164
   and IPCC 178–9, 180
   Kyoto Protocol 38, 165, 166,
     167
   Shell 55, 58, 151
   Texaco 99
global corporation 2, 17
goal attainment 159–60
Gore, Al 152, 164, 165, 166, 173
governance 199–200, 211–12
Grant, W. 162, 171
Greece 124
green consumerism 25, 105–6
green movement *see*
     environmental non-
     governmental organisations
Green Tax Commission (Norway)
     145
greenhouse gas (GHG) emissions
     197, 200
   corporate climate strategies 1,
     2, 15, 43–4, 66, 67, 69
   EU 125, 174
   ExxonMobil 48, 49, 51, 52,
     219
   government policy 116–17
   IPCC 178
   Netherlands 143
   Norway 129–30, 147, 153
   Shell 54–5, 56, 57, 58, 59
   Statoil 61–4, 65
   US 117–19, 120, 166, 173, 215
   see also carbon dioxide
     emissions; emissions trading;
     Kyoto Protocol
Greenhouse Gas Protocol
     Initiative 57
Greenpeace 121, 219
   Esso boycott 80, 112–13
   EU 171

Shell 80, 99–100
Statoil 64, 81, 82
US 111–12, 139, 170
Grubb, M. 160, 163

Herkströter, Cor 57
human resource availability 22,
     23, 98–9, 198

Ikwue, T. 162, 185
indifferent strategies 13, 14
influence 199, 207
   see also corporate influence;
     regime influence
Inglehart, R. 25, 26, 108
innovative strategies 13, 14
instrumental influence 35
interests 39
Intergovernmental Negotiating
     Committee (INC) 118, 129
Intergovernmental Panel on
     Climate Change (IPCC) 15,
     99, 176, 177–81, 191, 209
   US 118, 119
International Chamber of
     Commerce (ICC) 35, 61
International Energy Authority
     (IEA) 27, 122
international institutions 16, 17,
     206–7
International Petroleum Industry
     Environmental Conservation
     Association (IPIECA) 85
International Regime (IR) model
     8, 12, 13, 31–4, 41, 188–93,
     199, 200, 203–4, 206–7
   corporate climate strategies
     158–9
   corporate influence 34–7,
     159–76
   and DP model 208–9
   regime influence 18, 37–40,
     176–88
   two-level games 211
Ireland 124

Jänicke, M. 26, 29
Johannesburg Summit 219

Kasa, S. 130, 145
knowledge 39, 176–81, 191, 193,
    199, 204, 209
Kolk, A. 5, 23, 97, 98, 138, 180
Kyoto Protocol 39, 181, 189,
    197, 200, 217–18, 219
  corporate climate strategies 15,
    43, 66–7, 69
  corporate influence 163–8, 172,
    175, 189
  ENGOs 168, 170
  EU 182–4, 204
  ExxonMobil 48, 49, 52
  GCC 181–2
  and IPCC 119
  Netherlands 122, 124
  Norway 129–30
  in scenario planning 84–5
  Shell 54, 55, 56, 59, 101
  Statoil 62, 65
  US 111–12, 119–20, 152, 173,
    175–6, 190, 203–4, 215

leadership
  corporations 22, 23, 97–8, 198,
    205
  EU 218
Leggett, J. 6, 55, 93
Levy, D. L. 5, 6, 16, 23, 29, 97,
    98, 110–11, 139, 164, 180,
    212, 217
Levy, M. A. 5, 37–8
long-term agreements (LTAs)
    125–6, 127, 141, 142–3

McKinsey-derived matrix
    structure 53, 87–8
market opportunities *see*
    opportunities
Marrakech Accords 218
Marriott, P. 85, 86
Miles, E. D. 5, 32
MILJØSOK 141, 144, 146–7, 153
Ministry of Housing, Spatial
    Planning and the
    Environment (Netherlands)
    108, 121, 122, 123, 124, 186
Mobil 44, 210

  see also ExxonMobil
monitoring systems 20–1, 83–7,
    93–4
Moody-Stuart, Mark 97, 100
multinational corporations 2–7,
    16–18

National Environmental Policy
    Plans (NEPPs) (Netherlands)
    122, 123–4, 127, 141, 142
nationality 24
natural gas 19, 76, 77, 77
  Netherlands 127
  Norway 131–2
  Statoil 64–5, 113
natural gas liquids (NGL) 76, 77,
    77
Naturkraft 64–5, 82, 131–2
Naturvernforbundet 114, 145
Neale, A. 20, 53, 83, 84, 88
Nederlandse Aardolie
    Maatschappij 125
negotiated agreements 27, 140
neo-corporatism 140
neo-institutionalism 5
neo-realism 5
Netherlands 104–5, 202, 208
  climate policy 117, 121–5, 134,
    184, 192, 209, 214
  consensual approach 140–3
  DP model *148*, 149–51
  environmental attitudes 26
  policy instruments 125–8, *126*
  renewables 185–6
  social demand 106, 107–9, 110,
    115
Netherlands Employers'
    Association (VNO) 141, 142
New Game, The (Shell) 83, 84–5
Newell, P. N. 4, 5, 6, 35, 139,
    162, 163, 165, 168, 170, 212
Nigeria 80, 81, 105, 150
non-governmental organisations
    (NGOs) 6, 169
non-methane volatile organic
    compounds (NMVOC) 147,
    217
Norsk Hydro 64, 131, 145

Norsk Monitor 113–14
Norske Skog 64
NORSOK 146
Norway 104–5, 203, 208
    climate policy 117, 128–32,
        134–5, 209, 213
    corporate influence 217
    DP model *148*, 149, 152–3
    environmental attitudes 26
    national and international
        policy 35
    policy instruments 132–3
    political institutions 143–7
    social demand 106, 112,
        113–15
    see also Statoil
Norwegian Employers'
    Association (NHO) 145
Novem 126, *126*
numerical-democratic channel 29

offensive strategies *see* proactive
    strategies
Oil Pollution Act (US) 136
oil reserves 76, 77
Operations Integrity Management
    System (OIMS) 46, 47
opportunities 28–9, 33, 39–40,
    184, 191, 192, 193, 196–7,
    199, 214
    European emissions trading
        scheme 186–8
    green consumerism 105–6
    renewables 184–6
Organisation for Economic
    Cooperation and
    Development (OECD) 27, 28,
    110, 119, 208
organisational learning 20–2, 198,
    201, 207, 208
    company-specific factors 74,
        83–8, *89*, 90, *91*, 92, 93–4,
        95–6
organisational structure 21, 83
    company-specific factors 87–8,
        89, 90, *91*, 92, 93
Osherenko, G. 32, 39
Our Common Future 128

ownership 22, 23, 99–100, 149,
    198, 205

pattern matching 8
Pauly, L. W. 17, 23, 100
People Power (Shell) 83, 84–5
Petoro 61
Petroleum Act (Norway) 144
Pew Center on Global Climate
    Change 181–2, 216
policy instruments 117, 217
    Netherlands 125–8, *126*, 150,
        184
    Norway 132–3
    US 120–1, 133–4, 137
political institutions 29–31, 104,
    135, 149, 199, 206, 213–14
    conflict-oriented approach
        136–40, 152
    consensual approach 140–3,
        150
    mixed approach 143–7
political parties 108, 112, 114
Pollution Control Act (Norway)
    144
Portugal 124
Post, J. E. 3, 20
power 35, 39
President's Council on Sustainable
    Development 138
pressure 33, 39, 181–4, 191–2,
    193, 199, 213–14
proactive strategies 13, 14, 33,
    43–4, 192, 209
    CA model 20, 93, 96, 198
    DP model 26, 28, 116, 135, 199
    IR model 159, 183, 186, 199,
        215–16
    Shell 60, 66, 74, 158, 188, 197,
        201, 202
    Statoil 66
public voluntary agreements 27,
    28, 118–19, 137

qualified majority voting 36–7,
    174, 175

Ragin, C. 8–9, 22

Raustiala, K. 38, 164, 169
Raymond, Lee R. 50–1, 97, 100
reactive strategies 13, 14, 33
    CA model 96, 198
    DP model 117, 135, 152
    ExxonMobil 52, 66, 74, 158,
        188, 197, 201
    IR model 159, 183, 199,
        216–17
    Shell 60, 158, 188, 201
    Statoil 66
regime influence 5–6, 176, 191–2,
    193
    knowledge 176–81
    opportunities 184–8
    pressure 181–4
    snowball effect 217–18
regime perspective *see*
    International Regime model
regulatory capture 136
regulatory instruments 27, 28
regulatory pressure *see* pressure
Reich, S. 17, 23, 100
Reitan, M. 129, 145
renewables 1, 14, 98, 192, 207
    EU 204
    ExxonMobil 50–1, 80–1, 111
    Netherlands 123–4
    Norway 131, 133, 135, 153
    regime impact 184–6
    Shell 57–8, 59, 68, 101
    Statoil 64, 68–9, 114, 115
    US 133
reorientation in business areas *see*
    business reorientation
Retallack, S. 3, 4
Rondinelli, D. A. 3, 25, 28
Rosendal, K. 35
Rowlands, I. H. 4–5, 10, 16, 19,
    23, 24, 97
Royal Dutch/Shell Group of
    Companies *see* Shell
Royal Dutch Petroleum Company
    52
Russia 166, 218, 219

Saad, L. *109*, 112
Sand, P. H. 32, 33, 36

Santer, Ben 178–9
scenario planning 28, 86–7, 93–4,
    95–6, 101
    ExxonMobil 85–6
    Shell 83–5
    Statoil 86
Schenkel, W. 108, 143
science 191
    ExxonMobil 45–6, 48, 49–50,
        51
    international regime 176–81
    US 119
SDFI 60, 61, 76, 77, 77, 78
sea-level 121, 122
seven sisters 44, 52–3
shareholders 99–100, 205
Shell 9, 10, 52–3, 158, 188, 196,
    208–9
    Brent Spar 208
    CA model 94–6, *94*, 101
    capital availability 98
    change 210
    climate strategies 43–4, 54–60,
        66–70, *69*, 74, 200–1
    corporate influence 216
    DP model 105, 148, 149,
        150–1, 153, 202, 203
    emissions trading 183, 187
    environmental policy 53–4
    environmental reputation 79,
        80, 82
    environmental risk 75–8, 77
    GCC 167
    human resource availability 99
    Kyoto Protocol 163
    leadership 97
    monitoring of trends 83–7
    Netherlands climate policy
        121–2, 124
    organisational learning 90,
        93–4, 201
    organisational structure 87–8,
        89, 90
    ownership 23, 99–100
    policy instruments 121, 125,
        127–8, 134
    and political institutions 142,
        143

proactive strategies 197
renewables 185, 186, 192
scenario planning 28
science 176, 180, 191
social demand 105, 106, 107–9,
   110, 115, 116
Shell International 107, 134, 149
Shell International Renewables
   57–8, 59, 185, 192
Shell Netherlands 107
Shell Tradable Emission Permit
   Scheme (STEPS) 55, 187
sinks 169
Skea, J. 136, 162, 185
Skjærseth, J. B. 5, 6, 27, 33, 35,
   39, 40, 108, 113, 143, 160,
   162, 165, 167, 171, 174,
   186, 209
Skodvin, T. 5, 165, 167, 177, 186
Skolnikoff, E. B. 111, 119, 138
snowball effect 37, 38, 161, 191,
   217–18
SO$_2$ emissions 143
social demand 104, 105–6, 149,
   199, 205–6, 207, 213
   comparison 115–16
   Netherlands 150
   Norway 113–15, 153
   US 107–13, *109*, 151, 152
social demands 24–6
solar energy 64, 111
South Africa 80, 150
sovereignty 38, 191
Spain 124
Standard Oil 44, 109
State Pollution Control Authority
   (SFT) (Norway) 131, 144
state's direct financial interest
   (Norway) 60, 61, 76, 77, 77,
   78
Statkraft 64, 131
Statoil 9–10, 60–1, 196, 208
   CA model 94–6, *94*, 101
   capital availability 98
   climate strategies 43–4, 62–6,
   67–70, *69*, 200–1
   DP model 105, 148, 149,
   152–3, 203

energy taxes 213
environmental policy 61–2
environmental reputation 79,
   81–2
environmental risk 76–8, 77
intermediate strategies 197
monitoring of trends 86, 87
Naturkraft 131
Norway climate policy 128, 132
organisational learning 93
organisational structure 90, *92*
ownership 23, 99
policy instruments 132–3,
   134–5
and political institutions 144,
   145, 146–7
social demand 106, 113–15
Steger, U. 3, 4, 13, 18
strong influence 159
structural influence 35
Sununu, John 163

taxes *see* energy taxes
technological innovation 50–1,
   63, 119, 124
Texaco 99, 100, 167, 210, 212
tipping model 210
TotalFinaElf 212
tote-board diplomacy 37–8
two-level games 211

unanimity 36
uncertainty 28, 83
Underdal, A. 32, 34, 209
Union of Industrial and
   Employers' Confederation of
   Europe (UINCE) 161, 162,
   166–7, 184, 186, 187
United Kingdom 134, 162
United Nations Commission on
   Sustainable Development 128
United Nations Conference on
   Environment and
   Development (UNCED) 163
United Nations Conference on
   Human Development 26
United Nations Environment
   Programme (UNEP) 177

United Nations Framework
  Convention on Climate
  Change (UNFCCC) 1, 38, 39,
  164, 181, 217, 219
  corporate influence 160–3, 189
  ENGOs 168, 170
  and IPCC 119
  Norway 129
United Nations General Assembly
  177
United States 104–5, 202–3, 208
  auto industry 212
  climate policy 117–20, 133–4
  conflict-oriented approach
    136–40
  corporate influence 160, 172–3,
    175–6, 188, 214–15
  DP model *148*, 149, 151–2
  ENGOs 170, 172
  fossil-fuel lobby 164
  Kyoto Protocol 164–6, 167–8,
    181–2, 190, 203–4, 219
  oil industry 212
  policy instruments 120–1
  political institutions 214
  social demand 106, 109–13,
    109, 115–16
  two-level games 211

UNFCCC 162–3
United States Climate Action
  Report (USCAR) 118–19,
  120

Valen, H. 26, 113
Vattenfall AB 105–6
voluntary agreements 27, 28
  US 118–19, 120, 134, 137,
    151–2
VROM 108, 121, 122, 123, 124,
  186

Wallace, D. 136, 137, 142
weak influence 159
Weale, A. 26, 108, 122, 141
Wettestad, J. 39, 160, 183, 185
wind power 64, 114
World Business Council for
  Sustainable Development
  (WBCSD) 57, 219
World Meteorological
  Organisation (WMO) 177
World Resources Institute (WRI)
  57

Yin, R. K. 8, 9
Young, O. R. 32, 34, 37, 39

Lightning Source UK Ltd.
Milton Keynes UK
11 February 2010

149863UK00001B/9/P